TRANSFORMING MATTER

JOHNS HOPKINS

INTRODUCTORY STUDIES

IN THE HISTORY

OF SCIENCE

Mott T. Greene

and Sharon Kingsland

Series Editors

Transforming Matter

*A History of Chemistry
from Alchemy to the Buckyball*

Trevor H. Levere

THE JOHNS HOPKINS UNIVERSITY PRESS

BALTIMORE AND LONDON

© 2001 The Johns Hopkins University Press
All rights reserved. Published 2001
Printed in the United States of America on acid-free paper
9 8 7 6 5 4 3 2

The Johns Hopkins University Press
2715 North Charles Street
Baltimore, Maryland 21218-4363
www.press.jhu.edu

Library of Congress Cataloging-in-Publication Data

Levere, Trevor Harvey.
Transforming matter : a history of chemistry from alchemy to
the buckyball / Trevor H. Levere.
     p. cm. — (Johns Hopkins introductory studies in the
history of science)
   Includes bibliographical references and index.
   ISBN 0-8018-6609-X (acid-free paper) — ISBN 0-8018-6610-3
(pbk. : acid-free paper)

1. Chemistry—History.   I. Title.   II. Series.
QD11 .L45 2001
540′.9—dc21   00-011487

A catalog record for this book is available from the British Library.

Contents

Acknowledgments

This book owes much to the undergraduate and graduate students in lecture courses on the history of science and in seminars on the history of chemistry at the University of Toronto over many years of teaching. I am also much indebted to friends and colleagues. I would especially like to thank Lawrence Principe, William Brock, David Farrar, Anthony Fernandez, and Brian Baigrie for their comments on drafts of this book. I have cheerfully and gratefully adopted many of their suggestions. They have saved me from numerous errors, suggested the inclusion of several important topics, and helped me at many points to make arguments clearer. Juliana McCarthy at the Johns Hopkins University Press was most helpful and efficient at every stage of production.

Introduction

Chemistry today is fascinating and far ranging. We know something about the chemistry of the stars, distant crucibles where elements are formed. We know much about the chemistry of life—in biochemistry, complexity and richness are at their peak, supported by elegant and often simple concepts and models. There are just over a hundred kinds of chemical atoms, corresponding to the different kinds of chemical elements, but their possible and actual combinations are so many as to seem infinite. In only the past thirty years, the list of known compounds has grown by seven and a half million, which represents almost a tenfold increase in that short interval of time. Complexity, richness, and an economy of means give chemistry its intellectual appeal; utility and application, its universal relevance. Chemistry in medicine, agriculture, and industry and in its effects on the environment has transformed the conditions of life for our species, and for countless other species as well. Many of the necessities for our crowded planet are made possible by the applications of chemistry. Many of the problems that we have created while providing those necessities are also tied to chemistry, and so too will be their solutions. We make choices, perceive needs, and create social and political structures in which we use, wittingly and unwittingly, the science of our day.

Although chemistry has been important for millennia in its practical application to the needs and luxuries of human life, the discipline has not always been called chemistry. It has existed in very different forms, and in very different relations to neighboring sciences and crafts, in a flux that only accelerated as the years passed. The goals and concepts of a Chinese or Arabic alchemist of antiquity or the Middle Ages differed greatly from those of a chemist of the eighteenth-century Enlightenment, just as the aims and ideas of nineteenth-century research chemists were different from those of their predecessors and successors.

Chemistry has, historically, been in constant flux, both in its self-image and in relation to other disciplines that sought to co-opt or absorb it. It has had its sects, its rivalries, and many dead-ends. But from antiquity to the present, there have been men and women (formerly few women, now many) engaged in seeking to understand the way in which different substances are formed, how they react, and how they may be used. Chemistry is and has always been both science and craft or technique, depending on its material subject matter and on the tools and instruments developed for the manipulation and transformation of that matter. The chemistry of gases, for example, only became possible when apparatus were invented in the eighteenth century for containing and controlling gases. Chemistry, in its use of instruments and laboratory skills learned through experience, was and is a *practical* science, despite the existence today of the subdiscipline of theoretical chemistry.

But even chemists of the most practical sort could become lyrical in the face of the discoveries of their art. Was it not wonderfully strange that an inflammable gas and a gas supporting combustion and life could combine to produce pure water, showing no trace of its gaseous origin? So it seemed to many chemists around the time when the composition of water was discovered. How could substances with the same composition exist in different forms with different properties? That was a nice problem for the early nineteenth century. Chemistry has, however, been not only a science of ideas and discoveries but also marvelous in its uses. Think of modern medicine, which depends heavily on compounds synthesized in the chemical laboratory to fight disease.

The story in this book traces the study of the qualities and transformations of different kinds of matter from alchemical beginnings to the present. It follows a small number of themes: theories about the elements, the need to classify elements and compounds, the status of chemistry as a science, and the contributions of practice to theory. It explores these themes by concentrating on the contributions of some of the most influential and innovative practitioners of the science.

I have addressed the book to those seeking an introduction to the history of chemistry, whether they have a background in the science or not. Formulas are at a minimum, and concepts and technical terms are explained as they appear. A list of further reading appears after the final chapter. Chemistry in former ages will be as unfamiliar to most chemists today as modern chemistry is to nonchemists. There should be something familiar and much new for all but the specialist reader.

TRANSFORMING MATTER

1 First Steps

From Alchemy to Chemistry?

In 1980, scientists at the University of California at Berkeley used a particle accelerator to change an unimaginably small sample of bismuth into gold. It cost them $10,000 to make one-billionth of a cent's worth of gold. They showed that transmutation—the conversion of one chemical element into another—is possible today, but it is clearly not a paying proposition. At the beginning of the twentieth century, transmutation was not even within the reach of scientists. And earlier, in the nineteenth century, few chemists had any interest in such a crazy idea, which they regarded as part of a discredited alchemy. Most historians have had little interest in alchemy, except to show how wrongheaded and unscientific it was. Only a few eccentrics continued to search for the philosophers' stone, the fabulous substance that would change base metals into gold. Reputable chemists could not take them seriously.

But in the seventeenth century, alchemy still mattered. The seventeenth century is widely regarded as the age of the Scientific Revolution, the crucial epoch in the rise of modern science. Finding alchemy alive and well at such a time is surprising to those who see science as something essentially modern and alchemy as prescientific and misguided. Many seventeenth-century scientists and some politicians had a very different picture of alchemy. They could reasonably look forward to success with transmutation because their scientific theories could easily find room for it, and they had high hopes of economic as well as scientific benefit. True, no one had yet succeeded in the business of multiplication—the alchemical transformation of a little gold and a lot of base metals into a lot of gold—but several major figures thought a breakthrough was in sight. Robert Boyle (1627–91) was one such figure.

Today, we would call Boyle a scientist, but that word was not invented until the nineteenth century, when it was coined to describe practitioners of the sciences. *Scientist* by the mid-nineteenth century meant a person who studied one or more aspects of the natural world using the methods of chemistry, physics, and the other sciences. In the seventeenth century, those who studied the natural world tended to have broader horizons, deliberately including in their study metaphysical and even theological questions. These investigators saw themselves as natural philosophers, as, for example did Boyle and his contemporary Isaac Newton (1642–1727). Boyle was widely regarded as the leading English chemist of the seventeenth century. He was invited to accept the presidency of the Royal Society of London but declined because he refused to swear any oaths, including the president's oath of office. He used his influence at court to legalize the alchemical production of gold, which had been forbidden in England for nearly three hundred years. In the centuries when it had been illegal, the penalty for practicing alchemy was death. The crown would confiscate the property of a convicted alchemist so that any gold he might have made alchemically would go to the king. This was to avoid flooding the market with manufactured gold and thus destabilizing the economy. That no one before the twentieth century ever made any gold, alchemically or otherwise, had little effect on alchemists and their patrons—or on legal theory. When Boyle was able to have the law against multiplication repealed, the repeal, just like the old law, stated that any gold produced in the laboratory would be deposited in the royal exchequer. Kings had a strong interest in the alchemists' success.

Boyle has been often described as the father of modern chemistry, but if he believed in alchemy as science, then his chemistry must have been very different from ours. As we shall see, it was very different. Chemistry is not something that emerged from a prechemical past, to be defined once and for all. It is, as other historians have noted, the product of its own history and is constantly undergoing changes. Those changes make any definition a limited one—limited in time, in place, and in community. Boyle's chemistry was so different from ours that the author of a recent book about it insists on calling it *chymistry,* using the seventeenth-century spelling to emphasize its special nature. John Aubrey (1626–97), who was not a natural philosopher but was a keen observer and accurate recorder, described Boyle as a great alchemist.

Alchemy too can be seen as the continuously changing product of a history. It is time to identify its main themes and main variants.

Chinese and Arabian Melting Pots

The earliest alchemy that we know anything about was practiced in China by the fourth century B.C. It was interpreted through theories arising from Taoism, which was both a religion and a philosophy. In Taoism and the alchemy derived from it, the universe was seen in terms of opposites. There was an opposition between yang, the principle that corresponded to male, hot, and light, and yin, which was female, cool, and dark. This opposition generated five elements, also understood in terms of pairs of opposites: metal and wood, fire and water, as well as a central element, earth. Here was an explanatory model that could be applied to transformations and transmutations of material substances: change the proportions of the elements, and of their underlying principles, and you change the substances. But transmutation, including the transmutation of base metals into gold, was merely a tool for Chinese alchemy, a means to an end. Chinese alchemists were above all dedicated to discovering the elixir of life, which, when imbibed, would confer immortality—or at least prolong life indefinitely. The elixir was sometimes described as potable, that is, drinkable, gold.

Chinese alchemy was thus essentially medical in its goals. The multiplication of gold was a secondary affair, but it was nevertheless of interest. For once alchemists had discovered the secret of prolonging life, their next task would be to make the conditions of life comfortable, by generating wealth. Here multiplying gold would obviously help, since gold, then as now, meant wealth for its possessor. Because living longer and becoming wealthy were highly desirable, the alchemist-physician and the alchemist-multiplier were both worth employing, in the hope that they would succeed. That meant that serious and honest alchemists could find work, but so could cheats, frauds, and quacks. It did not take long before Chinese literature recognized these twofold divisions, between physicians and multipliers, honest inquirers and frauds.

There is one other distinction that needs to be made in considering Chinese alchemy: transformation of appearance versus transmutation of essence. (This distinction will be equally important when we come to Western alchemy and chemistry.) Jewelers and goldsmiths worked with precious stones and precious metals for wealthy customers. They also produced imitations of these expensive materials, which could, for example, be worn as costume jewelry. They were concerned with appearances, not with essences. Creating the external appearance of gold, whether by covering another metal with gold leaf or by some chemical process that brought gold to the surface of a mixture, required technical skills. These skills could be used openly, so that the purchaser knew he was buying only the appearance of gold, as today we may buy a piece of gold-plated jewelry. Or they could be applied fraudulently, with the intent of deceiving the purchaser, making him think that he had purchased solid gold.

Both fraudulent and honest work in Chinese alchemy, jewelry making, and metallurgy as well as in other kinds of applied chemistry (such as pharmacy) involved practical knowledge, important for alchemy and later for chemistry. The development of furnaces, the control of heat, the making of metallic alloys, the discovery of gunpowder, and the use of solvents all feature in Chinese alchemy. Alchemy remained important in China until the rise of Buddhism, which reached China in the first century A.D.

Was Chinese alchemy important for the emergence of Western alchemy? We do not know for certain, but it may have been. Trade routes and military conquest both involve two-way dissemination of ideas and techniques, and Eastern ideas may have come directly, or filtered through India, into the Greek, Roman, and Arabic worlds. Alexander the Great of Macedon and Greece was pushing eastward in his conquests at about the same time that Chinese alchemy was taking clear shape.

Alexandria, named for Alexander the Great, was a melting pot of cultures, technologies, and people, the intersection of trade routes, and the site of the greatest library the world had ever known. After classical Greece, it became the intellectual center of the Western world, drawing on the traditions of Greece, Egypt, Babylon, Rome, and beyond. Copper smelting had been achieved in the Bronze Age and was commonplace by the time that classical Greece was enjoying its extraordinary intellectual explosion. Metallurgy developed briskly in Greece as well as in Egypt and Babylon. Perfumes, cosmetics, dyestuffs, paints, and decorative pottery all involved the technical skills of applied chemistry—the manipulation, separation, combination, and modification of different substances.

Although these crafts could be practiced without reference to any theory of why they worked, from as early as the sixth century B.C., the ancient Greeks were astonishingly prone to asking why things were the way they were. They were, in short, natural philosophers, philosophers of science, with "science" meaning knowledge of nature. An important strand of Western alchemy began when the Greeks sought to account for empirical observations derived from metallurgical, cosmetic, and other crafts and techniques by constructing philosophical explanations for them. In the sixth century B.C., before the time of Socrates, Greek philosophers had argued that all substances were derived from an original prime matter. Somewhere around 450 B.C., Empedocles sought to explain the properties of matter and its changes by saying that all substances on earth consisted of four elements—air, earth, fire, and water—in different proportions. These elements were intimately mixed, like different colors of paint stirred together, rather than jumbled, like bricks in a heap.

A century later Aristotle (384–322 B.C.), who had been a pupil of Plato and tutor to Alexander the Great, took up the idea of prime matter and Empedocles' four elements, and he added four qualities, hot, cold, wet, and dry. Qualities imposed on prime matter generated elements, which, when mixed, constituted the substances that we find in and on the earth. Earth was cold and dry; water, cold and wet; air, hot and wet; and fire was hot and dry. The pure, or philosophical, elements were never found in isolation, but always mixed with more or less of the other elements. Given this scheme, how were metals formed? Miners found them in the earth, where a natural process had imposed the qualities of metals on the original prime matter. Different metals represented different degrees of acceptance of these qualities. The process could be compared to the biological sequence of conception, pregnancy, and birth: the earth gave birth to metals, which grew in its womb.

We have considered one approach to alchemical theory in classical Greece. By Hellenistic Alexandrian times, alchemists had come to believe that they could replicate in the laboratory the process of the growth of metals and that they could also accelerate it. The goal became to strip metals of their properties, thus reducing them to prime matter, and then to impose the qualities of gold on the resulting undifferentiated mass. As alchemy took shape, the power of imposing qualities was believed to reside in what became known as the philosophers' stone. The process could be tracked by the color of the substances that the alchemist produced, with the color sequence revealing the operation's success or failure. The desired sequence went from lead to a black substance, because black represented the absence of color and was thus appropriate for prime matter. The next colors, in order, were white, yellow, and purple. Goldsmiths had long been interested in the colors of different alloys. Now their skills and experience were useful to alchemists.

Two other lines of thought are especially important for the development of Alexandrian alchemy. These are astrology, which had been established in Babylon before Greece, and Stoic philosophy. Stoic philosophy and Babylonian astrology both posited a cosmos governed by correspondences between great and small, so that what happened locally on earth reflected larger patterns in the cosmos. From this belief there emerged a detailed account of the correspondence of seven metals to the seven known planets, which in the ancient world included the sun and the moon. Gold was matched with the sun, silver with the moon, iron with Mars, mercury with the planet Mercury, copper with Venus, tin with Jupiter, and lead with Saturn. Alchemy, in the light of astrology, depended on the influences of the planets. Stoics added the notion of a world soul or spirit.

It was of course not necessary to be an astrologer in order to be an alchemist. One of the leading Alexandrian alchemists, Zosimus, who lived in the fourth century A.D., was clearly skilled in laboratory manipulations. He was a practical alchemist who knew a lot about distillation, sublimation (converting a solid directly into a vapor by heating it), filtration, the use of furnaces, and more. Zosimus and his contemporaries knew about contemporary techniques in metallurgy, dyeing, glass making, and other applied crafts. They have left us descriptions of apparatus designed and used for alchemical purposes, including a variety of stills and condensers, as well as furnaces, water baths, and sand baths (see Chapter 2). Much of the apparatus used by the Alexandrians was still in use in essentially the same form well into the Middle Ages. Many of the techniques of alchemy and later of chemistry were developed and elaborated in Alexandria at the time of the Roman Empire. But Zosimus was more than a skilled practical chemist. He also claimed to possess the key to transmutation—the philosophers' stone of later alchemists.

From *Arabian Nights* to *Canterbury Tales*

After the Islamic Arabs took Alexandria in the seventh century A.D., the center of learning shifted to Damascus and Baghdad, where renewed growth in alchemy came along with cultural resurgence under the new religion of Islam. Alchemical texts were translated from Greek into Arabic beginning in the eighth century. Under the patronage of Harun al-Rashid, who is best known today as the caliph in the tales of the *Arabian Nights,* scholars translated Hellenistic alchemical tracts into Arabic. Later scholars in Christian Europe attributed some of these translations, other original alchemical manuscripts, and numerous technical alchemical skills to Jabir ibn Hayyan, who is said to have lived from around 721 to around 815 and to have been court physician to the caliph. Unfortunately, it is probable that Jabir never existed. It was convenient, however, for later medieval scholars to attribute both writings and technical advances in alchemy to a distinguished predecessor, even an invented one. I shall write of him as if he existed, but I shall place his name in quotes as a reminder of his status, which is at best hypothetical.

"Jabir," or at least the school associated with him, made numerous contributions to laboratory practice, including refined techniques of distillation, the preparation of medicines, and the production of salts. The determination of Arabic alchemists to find the constituents of chemical substances led them to the discovery and use of strong reagents, chemically active substances used to test for the presence of a variety of other substances. They also developed theories to account for the action of those reagents. Acids, for example, could corrode a metal, a process that the alchemists interpreted as the separation of the metal into its constituents. When those constituents had been reduced to their elements, they were expected to work powerfully in producing the agents of transformation. If this were so, then analysis and subsequent synthesis could contribute to the discovery of the philosophers' stone, sometimes known as the "elixir."

(Like Aristotle, "Jabir" believed that metals grew in the earth. Aristotle had adopted Empedocles' four elements, but he interpreted the birth of metals to the combination of a wet and a dry exhalation arising from the earth under the influence of the heat of the sun. Following Aristotle, Arabic alchemists distinguished philosophical elements or principles from the substances of everyday experience. "Jabir" linked Aristotle's wet and dry exhalations with philosophical Mercury and Sulfur, which were different from and purer than the mercury and sulfur of the laboratory. By purifying everyday mercury and sulfur and appropriately adjusting their proportions, the alchemist could make gold. The making of gold and the extension of life were both important in Arabic alchemy.)

(Theory and practice were closely entwined for the Arabic alchemists of the eighth and ninth centuries. They asserted that every substance contained its opposite, in a hidden, or occult, way. Silver was cold and dry externally, but hot and wet internally. Gold was hot and wet externally, but cold and dry internally. In order to make gold one therefore needed to exchange the internal and external qualities of silver. |

Medicine, metallurgy, and all the applied arts involving chemistry and alchemy thrived in the first centuries of Islam. So did the armies of Islam, sweeping from the Middle and Near East westward across North Africa and then northward to conquer the Iberian Peninsula (now Spain and Portugal). Spain under Islamic rule saw a flourishing of Jewish scholars and at first a tolerance for Christian scholars, alchemists among them. It was through Spain and its Arabs, gradually reconquered by Christian forces, that Greek, Alexandrian, and Arabic alchemy made their way into the Christian West.

Christian scholars translated Greek and Arabic texts into Latin. Their translations were often accompanied by a good deal of revision and modification. A thirteenth-century Italian Franciscan alchemist called himself Geber, the Latin version of "Jabir," to take advantage of "Jabir"'s august reputation. The works of Geber were indeed based on Arabic alchemy, but with significant modifications. Geber's system included a kind of corpuscularianism grafted onto the original stock, so to speak, such that inner and outer qualities could be reinterpreted in terms of inner and outer layers of minute particles, or corpuscles. Corpuscularianism is like atomism, but with one crucial difference: atoms, by definition, are indivisible, or at least they were believed to be indivisible prior to the discovery of subatomic particles in the modern age.

Geber dealt with corpuscles that could in principle be divided. His mercury could penetrate into metals and modify their inner structure, a step on the way to the transmutative production of gold.

Geber's writings also indicated that alchemical success was God-given, an attitude that reinforced the spiritual aspects of medieval alchemy. The quest for the philosophers' stone came to be seen as a metaphor for the salvation of the soul, and alchemical imagery increasingly used Christian metaphors, including references to death and rebirth. The destruction of qualities—for example, "stripping" lead of its metallic properties—corresponded to death; the alchemical production of gold was then a kind of resurrection. The oppositions that Arabic alchemists had written about (e.g., the opposition between philosophical Sulfur and philosophical Mercury) were also described in sexual terms. It is often difficult to know whether the imagery of alchemy was merely metaphorical or the correspondences implied were seen as real. Meanwhile, laboratory practice advanced. Alchemists extended their knowledge and classification of salts and produced stronger corrosive solvents, first nitric acid, then hydrochloric acid, and finally sulfuric acid.

Even before alchemy entered the Latin West, Arabic alchemists and physicians did not all agree about the possibility of transmutation. From the tenth century on there was a lively debate, with some following "Jabir"'s lead and others insisting that metals were natural species, just like animals and plants, and were not interconvertible or transmutable. The debates persisted when alchemy spread into Christian Europe. It is not surprising that in the following centuries, alchemy maintained its mixed reputation and wide embrace, encompassing both attempts at transmutation and what we would recognize as practical science (including metallurgy and, increasingly, chemical medicine).

Literary and popular satire took gleeful aim at fraudulent alchemists. A splendid portrait of such a character figures in one of Chaucer's *Canterbury Tales*, "The Canon's Yeoman's Tale." The narrator is servant to a shabby, unsuccessful, and imaginatively fraudulent alchemist. If the canon's claims were to be believed, he could use his alchemy to take the pilgrims' route to Canterbury "and pave it all with silver and with gold!" But when he hears that his servant is about to tell all, he makes a hasty departure. The servant decides that he is finished with his master and does indeed tell all. With a stained and sooty face, the result of endlessly blowing into the fire to keep it going, he explains that their goal was to learn to multiply, to turn a little gold into a lot. Failure in this enterprise never stopped them from tricking gullible and greedy patrons:

> But there are lots of folks that we take in,
> And borrow gold from—say a pound or two,
> Or ten, or twelve, or many times that sum,
> And make them think the very least we'll do
> Is double the amount: make one pound two.*

Hide a little gold or silver up your sleeve, or conceal it in a hollow rod sealed with wax that melts on warming. Then introduce the gold or silver into a crucible when the victim-to-be is not looking, and tell him that the metal was produced by multiplication. Once he is persuaded that he has seen multiplication at work, he will readily come forward with his own gold or silver, for further multiplication or transmutation. Then take the money and run. By such tricks, the canon conned his victims into supporting his alchemical endeavors. But even in the case of such a cheat, it is important to recognize that he used fraud to finance earnest attempts to discover the philosophers' stone.

The yeoman's tale includes a good deal of technical information, making it clear that the range of substances available to the alchemist had expanded greatly since Alexandrian times. He tells of the importance of correct proportion by weight of substances (including the bright golden-yellow arsenical pigment called orpiment) which are ground to a powder, put into an earthen pot, and luted, that is, sealed with clay or cement to make sure that nothing, not even air, escapes. He notes the different degrees of heat used and describes processes such as amalgamation (the softening of metals by combining them with mercury, or the union of two or more metals into an alloy), calcination (applying a roasting heat to nonfusible substances), and sublimation. He runs through a list of apparatus made of earthenware and glass, much of it for distillation. He recites the names of salts, other minerals, herbs, acids, solvents, and "divers powders, ashes, dung, piss, and clay." And he recites the correspondence of the names of planets with different metals, a correspondence that popular belief in astrology had made well known by Chaucer's day.)

Paracelsus: Nature as Alchemist

Alchemy saw significant advances in the range of techniques and the knowledge of mineral and plant substances from the fourteenth to the sixteenth century. There was also the addition of a kind of atomism to the basic theory and a growing emphasis on the multiplication of gold instead of medical alchemy.

*Geoffrey Chaucer, *The Canterbury Tales,* trans. David Wright (Oxford: Oxford University Press, 1985), 427–50, at 430.

These changes were not seen as revolutionary. But in the sixteenth century there came an individual who had no interest in having his works attributed to past alchemists or physicians, as he was sure he was better than all of them. His unwieldy name was Theophrastus Bombastus von Hohenheim (1493–1541), but he called himself Paracelsus as a way of claiming superiority to the great Roman physician Celsus (para-, or παρα-, is Greek for past or beyond).

He created a revolution in alchemy and in medicine, which for him were two intimately related disciplines. He had little interest in making gold by multiplication or other transmutative processes. For him, alchemy was most valuable, perhaps only valuable, when it was used in the service of medicine. He rejected the mercury-sulfur theory that Arab alchemists had grafted onto Aristotle's wet and dry exhalations. This theory of two elements (a dyad) was inadequate for Paracelsus. He saw sicknesses as distinct and specific, arguing that distinct and specific medicines were needed to cure them. This was in complete revolt against the reigning medical theories of his time. These theories, following the Greek physician Galen, posited four humors akin to Aristotle's four elements and saw sickness as the result of humoral imbalance, thus medicine operated mainly by adjusting the imbalance through bleeding, purging, and similar debilitating treatments. The treatments, like the overall perception of illness, were systemic and general, not specific and local. Paracelsus aimed to overthrow traditional medicine along with traditional alchemy. He was a revolutionary and an iconoclast.

If alchemy was a servant for Paracelsus, it was also a glorious and powerful one. He saw the transformation of the invisible into the visible as essentially alchemical, and so regarded the processes of living nature as alchemical. The growth of animals and plants, the ripening of fruit and vegetables, the processes of fermentation in making beer and wine, the digestion of food, indeed all natural processes involving transmutation, growth, and development were alchemical. Nature was an alchemist; and God, who ruled Nature, was the supreme alchemist.

Medieval alchemists had generally adhered to a dyad theory, in which Sulfur and Mercury were the principles of all metals and change was produced by the interaction of these two principles. Substances rich in Sulfur were more combustible, while those rich in Mercury were less so. Paracelsus took this dyad theory and added a third principle of Salt to it. His three principles—the *tria prima,* or three first things—were able to explain the alchemical transformations of all bodies. This material trinity matched the Holy Trinity in heaven as well as the three principles of which we are made: vital spirit, soul, and body. Important in this scheme are correspondences between the great world, the *macrocosm,* and the little world, or *microcosm,* of our bodies, between heaven and earth, and between alchemical processes in the laboratory and physiological processes in the human body. The overall theory, like its predecessors in alchemy, was of exceeding generality. But when it came to medicine, Paracelsus got down to particulars.

He looked for specific causes and symptoms of distinct diseases. He worked in mining regions of German-speaking lands, and one of his achievements was to identify both a lung disease of miners (silicosis) and its cause in miners. Where treatment involved medicine, alchemy was essential for Paracelsus. It was the science of the preparation of medicines.

Paracelsus was the founder of medical chemistry, known as *iatrochemistry,* to which we shall return in Chapter 3. After him came a Flemish alchemical physician, or iatrochemist, Jan Baptista van Helmont (1579–1644). Van Helmont combined Paracelsus's organic, or biological, model with the corpuscularianism of Geber. He used Genesis, the first book of the bible, as a guide to his matter theory. There was no element of fire in Genesis, so he rejected it as an element. The bible tells of water *above* the firmament, which for him meant that heaven and water were made prior to earth. He concluded that there were two original elements, air and water. Water was the primordial substance, within which sulfur and mercury somehow form distinct but inseparable parts. The mercury and sulfur could be rearranged, and so could mask or change the appearance of water. This process of masking was at the heart of Van Helmont's most famous experiment, in which he potted a willow sapling with some earth and weighed it. He watered it regularly over a period, weighed it again, and then deduced that the increase in weight and the growth of the sapling had to come from water, since it had not come from the soil and water was the only substance that had been added.

Looking Both Ways: Isaac Newton

When Robert Boyle began his work, alchemy was alive and well. It was widely although far from universally used in medicine. The search for the philosophers' stone was still taken seriously by leading natural philosophers, and many kings and princes were keen to have their own alchemists. Corpuscularianism was also alive and well, and Boyle was able to combine the latest brand of corpuscularianism with alchemy, to powerful effect. He did not accept Paracelsus's element theory, nor was he keen on Van Helmont's; and he was not much impressed by the way in which seventeenth-century iatrochemists had added phlegm and earth to Paracelsus's *tria prima.* But he did believe that transmutation was possible and that the alchemical production of gold by multiplication was a reasonable prospect. He was willing to put his money where his mouth was and to pay for knowledge as well as to seek it in the laboratory.

Boyle was also a close colleague of another natural philosopher, who would come to have even greater distinction than he, Isaac Newton (1642–1727). Newton's crowning achievement was the elucidation of the law of gravitation and its application to celestial and terrestrial phenomena. He was a professor of mathematics at Cambridge University. He was also for many years president of the Royal Society of London and, more briefly, a member of Parliament and Master of the Mint. He was, in short, the very model of a modern major scientist and statesman of science. Until recently, historians have accepted that strict model and been reluctant to recognize that he was also a serious student and practitioner of alchemy. It is arguable that alchemy was as important to him as mathematical physics and astronomy. Newton and the age in which he lived were clearly more complex than the old historical model perceived.

Newton was engaged in alchemy for more than forty years. These years spanned the writing of his two great books, *The Principia: Mathematical Principles of Natural Philosophy* (first edition 1687), and *Opticks* (first edition 1704). He studied the literature of alchemy and was profoundly absorbed in its experimental practice, so much so that he has been well described as a "philosopher by fire." Newton, both in his accounts of universal gravitation and in his pursuit of alchemical transformation and transmutation, talks about God and discusses active principles, the tools of divine activity in the world. The God-grounded unity of truth meant for Newton that all avenues to truth, including alchemical wisdom and experiment, were mutually reinforcing.

He was a corpuscular philosopher. Early on, Newton became convinced that matter came from a single root, such that there was a unity of matter. That made transmutation possible in principle. As he wrote in the first edition of his *Principia,* "Every body can be transformed into a body of any other kind and successively take on all the intermediate degrees of qualities."* The question was how to effect such transformations. He looked to a universal vegetative principle, a material spirit and source of activity that would generate gold from a metallic seed or semen. He sought the substance that would best embody that principle, which, when combined with the action of fire, would first reduce a substance to chaos, like the prime matter of the Greeks, and would then move on to generation.

As a corpuscular philosopher, Newton was able to echo Geber, with inner and outer qualities of matter, and to explain transmutation as the result of

---

*Isaac Newton, *The Principia: Mathematical Principles of Natural Philosophy,* trans. I. Bernard Cohen and Anne Whitman (Berkeley and Los Angeles: University of California Press, 1999), 795.

changes in the inner arrangement of matter and of qualities. Mere changes in arrangement corresponded to mechanical chemistry, while a living vegetative spirit, the alchemical carrier of divine activity in the world, produced more profound changes. Newton's God was an active God, forever working in nature, and was in part an alchemist. This spiritual dimension of alchemy encouraged Newton to be cautious and even secretive when it came to transmutation. With his laboratory in Cambridge where he spent so many hours in the experimental pursuit of alchemical wisdom, Newton looked back to the great alchemists of the past as his guides.

Alchemy began in antiquity, as a practical and a theoretical or philosophical pursuit. Combining theory and practice, it has to be taken seriously as science. Its fundamental theories were wrong, but that does not rule it out as science. Most old science turns out to have been wrong, including much of Newton's work on optics. Einstein's relativity has overtaken Newton's mechanics. But Newton's science was still good science, providing a framework and a tool for the acquisition and organization of knowledge. Much of our science today will, sooner or later, be old science, and future scientists will reject much of it.

Alchemists, over a period of hundreds of years, devised apparatus and instruments as well as techniques for using them that remained valuable even after alchemical theories had been discredited. They constantly extended the range of known chemical substances and developed criteria for identifying and classifying those substances. They developed and refined notions of purity (as we shall see, the question of the purity or impurity of reactants has a major role to play in the history of chemistry). Not least among the achievements of alchemists was their success in establishing their science upon a succession of theoretical foundations. It was possible, at least until the end of the seventeenth century, to integrate alchemy with major trends in the growth of contemporary science, such as medical chemistry and mechanical and corpuscular philosophy. It is worth noting both continuity and discontinuity in the history of alchemy. As we have seen, here was a science that underwent more than one revolution of its own before the Scientific Revolution of the seventeenth century. As will be clear from the example of Robert Boyle, the leading English practitioner of alchemy, chymistry, and chemistry during the Scientific Revolution, it is simply not possible to make a sharp separation between alchemy and chemistry in the seventeenth century.

2 Robert Boyle

Chemistry and Experiment

Robert Boyle (1627–91) was a famous seventeenth-century chemist. His contemporary John Aubrey wrote of him: "His greatest delight is chemistry. He has at his sister's a noble laboratory and several servants (apprentices to him) to look [after] it. He is charitable to ingenious men that are in want, and foreign chymists have had large proof of his bounty, for he will not spare for cost to get any rare secret."* He certainly learned from others, notably the American alchemist George Starkey (1628–65), who was born in Bermuda, was educated at Harvard, spent some fifteen years in London, and died in the Great Plague. Although Boyle has often been referred to as the father of modern chemistry, his chemistry was not modern chemistry. He was an alchemist as much as a chemist. Modern chemistry rests in part on an idea of chemical elements developed in the late eighteenth and early nineteenth centuries, and Boyle by no means anticipated that idea. The origins of chemistry are too complex for any individual to be given sole credit. Boyle is, however, an important figure in the history of chemistry, and in two respects he deserves at least some credit as the founder of modern chemistry.

His major contributions were twofold. First, he convincingly championed chemistry as an important part of the new natural philosophy of the seventeenth century. More precisely, Boyle argued that chemical philosophy and corpuscular philosophy provided important support for one another. We will soon consider the nature of corpuscular philosophy. For now, it is enough to note that it offered mechanical explanations, based on the behavior of corpuscles. These corpuscles might be aggregations, groups, or clumps of atoms, which were in principle divisible. Alternatively, they might simply be individual atoms, which by definition were indivisible. Boyle made chemistry compatible with the new, fashionable, and dominant kind of scientific explanation. His second major contribution, partly borrowed from Starkey, was the development of an *experimental method* in chemistry that made it fit into the new

*John Aubrey, *Brief Lives*, ed. R. Barber (London: The Folio Society, 1975), 55.

authoritative public practice of science championed by the Royal Society of London. It is a pleasing coincidence that his best-known chemical book, *The Sceptical Chymist*, was published in 1661, one year after the original foundation of the Royal Society.

Boyle was born in English-occupied Ireland, the seventh son of one of the richest men in the kingdom. When he was eleven years old, he went on a European tour, in what was to become a tradition for the sons of the wealthy. It is unfortunate that he was so young—too young to participate in the latest exciting philosophical debates. Paris at that time was the European center for corpuscular philosophy. Pierre Gassendi, an ordained priest, natural philosopher, atomist, and friend of Galileo and Kepler, was at the center of what was essentially an unofficial university for the study of mechanical and corpuscular philosophy. There a variety of different forms of atomism and mechanism jostled with one another(René Descartes invented his own brand of particle theory, in which space was considered to be completely filled with particles of different sizes. Boyle later learned much about these philosophies, and it is useful to say more about them now and about their origins.)

In fifth century B.C., the Greek philosopher Leucippus invented a cosmology in which the world was described as made up of atoms moving in empty space. Democritus later took up and developed Leucippus's theory. He portrayed a world in which an infinite number of eternally unchanging atoms moved in a vacuum, and, through their chance combinations, produced all the different bodies in the world and accounted for their qualities.)Aristotle, in the fourth century B.C., was extremely disturbed by such views. His reasons for being disturbed were good ones. For Aristotle, explanation was above all an account of the *causes* of things and of the *purposes* that governed them. In a world made up of atoms moving at random and governed only by chance, how could one talk of cause and purpose? How could one really explain anything? There were other problems too. Aristotle believed in continuity, whereas atoms moving in a vacuum inevitably introduced physical discontinuities. He did not dismiss all kinds of atomism. He believed in the existence of minute parts of matter which became known as *minima naturalia,* but these were in his view neither eternally unchanging nor existing in a vacuum, and in spite of being called *minima* (i.e., "least things"), they were not even indivisible. It was Democritus's doctrine of atoms and the void that Aristotle found philosophically intolerable, and he therefore rejected it.

That doctrine did, however, find some favor after Aristotle. Epicurus, another Greek philosopher, developed a philosophical system with cosmological and ethical components. The cosmology followed Democritus. The ethics found value in a world governed by chance, a godless world, in which pleasure was the only good. Epicurus's ideas survived into Roman times, when the poet and philosopher Lucretius wrote six books of verse called *On the Nature of Things* to popularize them. Lucretius, like Epicurus, saw atheism as a positive feature of atomism.

When Greek and Roman learning next surfaced in the Western world, it followed the same paths as alchemy had, by way of Alexandria into Arab civilization, which eventually extended into Spain. Following Spain's reconquest by Christian forces, its treasure house of Greek, Roman, and Arabic learning passed into the Christian West. Philosophical learning, including natural philosophy, became an avocation for churchmen, who pursued knowledge in the newly founded universities and in church establishments. Aristotle, suitably interpreted, was favored by them, and, since they did not have access to the works of the atomists, they did not have to wrestle with doctrines of atoms and the void.

Then came the Renaissance, a period of the recovery of ancient learning and of an unstoppable flow of new observations and new ideas, often emerging from or inspired by the old. Lucretius was rediscovered, and so was Epicurus. Greek atomism became fashionable at the French court. But just as Aristotle in the twelfth and thirteenth centuries had had to be interpreted and modified so as to be reconciled with Christianity, so too did atomism in the seventeenth century. Gassendi undertook the Christianization of atomism. Atoms, he explained, were not eternal but created by God. Their movement in the void was not random but the result of their God-given initial motions, which made them agents of divine purpose.

A work justifying atomism in ways very much like Gassendi's was also sought in England. A very unreadable but successful one appeared, written by Walter Charleton. By the time the Royal Society of London was founded, "mechanism" of one kind or another was the new orthodoxy, and both atoms and corpuscles were fitted into the world.

Boyle came to be an adherent of corpuscular philosophy. Aristotle's philosophy, with its four terrestrial elements occupying a plenum, was incompatible with a doctrine of atoms and the void, or of a universal matter. Boyle subjected Aristotle's theories, as far as they applied to chemistry, to serious criticism. He did the same to Paracelsus's theories, which were based upon three elements, the *tria prima*. Boyle's criticisms were both rational and experimental in character. They did not prevent him from pursuing alchemy as part of his chemistry. It is this unique combination of what were later separated into the twin pursuits of chemistry and alchemy, to the latter's disadvantage, that has led some historians to characterize Boyle's avocation as "chymistry," to distinguish it from what came before and after. I shall continue to use the spelling "chemistry" and the words *chemistry* and *alchemy*. It is, however, important to realize that Boyle's chemistry is very different from that of the late eighteenth century, just as that chemistry is very different from today's. The idea that alchemy involved metallic transformation while chemistry had other goals was not widely accepted until the very end of the seventeenth century. That acceptance was based on a mistake in historiography, which led to an incorrectly narrow interpretation of the range of alchemy. Chemistry at any given time is the product of a continuing history, subject both to evolution and on occasion to revolution.

A Christian Alchemist

When Aubrey wrote that Boyle spared no expense to learn rare secrets, he was referring especially to alchemical lore, which was frequently kept secret from all but adepts. Boyle's hostility toward Paracelsus's three-principle theory and toward Aristotle's four elements did not mean that he was against the philosophical theory of Sulfur and Mercury associated with Geber. Indeed, he believed that real, true, or philosophical Mercuries and Sulfurs could be isolated from metals. What he would not do was extend this idea to all substances. Nor did he accept the idea that the separated Mercuries and Sulfurs were elemental. He went so far as to believe that he had himself prepared the sulfur of gold and had been shown its mercury. He worked to animate mercury to produce more noble Mercuries, and he believed that he had succeeded in the alchemical degradation of gold to silver. If that was possible, so too was the reverse process, the transmutation of silver into gold.

Boyle had a lifelong interest in philosophical Mercury, which he thought would dissolve the metals into their constituents. He believed that the philosophers' stone existed and that it could serve as a universal medicine. He also believed that it was spiritually active and could facilitate communication with angels and rational spirits. Because God had created the world, seeking to understand the causes of natural phenomena was a path to the understanding of God's work. Natural philosophy, including Boyle's chemistry, was therefore a religious, indeed a Christian, activity. Boyle, by constructing a union of his religion with the fundamental ideas of gold making, showed how deeply he was committed to the enterprise.

Boyle accepted the possibility of transmutation, and he distinguished between transmutation performed with the philosophers' stone and other modes of transmutation. The philosophers' stone could transmute any metal into gold, and it could indefinitely, perhaps endlessly, multiply itself and the gold that it produced. Transmutation could also be explained simply by the re-arrangement of the corpuscles that made up a particular body, and it did not work on most metals. Use of the philosophers' stone could legitimately be re-served for initiates, and secrecy could be justified in its pursuit and practice. In contrast, other aspects of Boyle's chemistry needed to be open to scrutiny and criticism. Boyle sought to reconcile his science with the new and respectable corpuscular philosophy, on which he based his careful construction of a new experimental method.

The Philosophy of Experiment

Boyle devoted his life to developing details of a new way of knowing. He called this way the experimental philosophy. His experimental philosophy was his own, but he built it on foundations established by others. He did not see as far as his young friend Isaac Newton, but he could have said, as Newton did, that if he saw further than other men had, it was by standing on the shoulders of giants.

Like many of his contemporaries, Boyle found experimentalism a satisfy-ing alternative to the sterility of Aristotelian learning. The fiercest and most influential critic of that learning in England in the generation before Boyle's was Francis Bacon (1561–1626), lawyer, statesman, and philosopher. Bacon wanted to found his theories on reliable information rather than on specula-tion or tradition. He believed that systematic and comprehensive natural his-tories—bodies of information about nature derived from experience rather than from the authority of ancient books—would provide the proper founda-tion. One did not understand what happened in nature by witnessing a single event or performing a single experiment. Knowledge needed lots of informa-tion as its foundation, and phenomena first needed to be observed before one worried about their causes.

Bacon was a lawyer, and experimentation for him was a way of putting na-ture on trial and making it reveal its hidden workings, the causes of phenom-ena. Judge and jury in a court of law needed a body of evidence to decide where the truth lay, and they needed experience and common sense in judging that evidence. Bacon argued that the natural philosopher worked in the same way, requiring lots of information to reach a conclusion. Bacon, like Boyle after him, was interested in searching out "the more subtle changes of form in the parts of coarser substances." In experiments, the natural philosopher, like the lawyer, should use his sense "only to judge of the experiment, and . . . the ex-periment itself shall judge of the thing." The point of collecting data and organizing them into systematic data banks, or natural histories, was "to give light to the discovery of causes."* Philosophers were to use experiment and sense to analyze and dissect nature. Chemical analysis offered one way of dissecting nature.

Boyle followed Bacon's approach closely here. He saw chemistry as a tool for probing nature, and he argued that other natural philosophers (including the chemists whose theories he criticized) had made improper use of empirical evidence. They had arrived at general conclusions and theories by reasoning with insufficient data. Sometimes they were so unwise as to derive a theory from just one experiment. Boyle, like Bacon, wanted his database to be complete. That was a huge enterprise, one that even if ultimately achievable would at best be out of reach for a very long time. Conclusions reached in the meantime were therefore necessarily tentative; they were hypotheses, which could be overthrown if counterevidence later came to light.

Boyle was a great admirer of those who showed the most ingenuity and rigor in their experiments. Remember that experiments were not merely a matter of observing; they were a way of putting nature on trial, in order to understand its causes and inner workings. Boyle particularly admired Galileo as an experimenter. Freely falling bodies were hard to observe, so Galileo had hit on the strategy of slowing them by letting them roll down inclined planes and then reasoning about what he observed. Boyle thought that chemists too needed to show how experiments were done, to bring experiments "out of their dark and smoky laboratories" into the light of informed public scrutiny.†

Distinguishing between observation and experiment, Boyle believed that chemists made observations "of what nature does, without being over-ruled by the power and skill of man." These observations were building blocks in the database of natural history. Experiment, in contrast, involved intervention, "when nature is guided, and as it were, mastered by art."‡ He realized that experiments went beyond the outward appearance of things. They depended on some prior theoretical knowledge, which was itself based on an interpretation of the history of nature. Theory and experiment therefore relied on one another.

To succeed, experiments needed to be reproducible, both by the chemist who originated them and by other natural philosophers. Boyle gave careful at-

---

*Quotations in this paragraph are from Bacon's works, as quoted by Rose-Mary Sargent, *The Diffident Naturalist: Robert Boyle and the Philosophy of Experiment* (Chicago: University of Chicago Press, 1995), 51.

†Boyle quoted in Sargent, *The Diffident Naturalist,* 71.

‡Ibid., 71, 137.

tention to the circumstances that could frustrate that need. One of the most important considerations was that chemists had to work with pure substances, and tests were required to ensure that purity. Impurities could come from the original source, for example, in complex minerals. They could be introduced deliberately, by fraud. They could creep in over time, as in the process whereby wine turns to vinegar. And there were lots of practical considerations. The scale on which an experiment was carried out could be a problem. Consider that fire would not work uniformly throughout a large sample but would work evenly in a small sample. Instrument quality was crucial. For example, if the connections between vessels were inadequately luted (sealed), reactants would escape. Even if such leakage did not change the process being investigated, it meant that precise quantitative measurement was impossible; and quantification was something that Boyle was keen on.

Chemistry and the rest of natural philosophy offered a way of learning about God's activity in the world and were therefore religious activities. Boyle was confident that God would gradually reveal all knowledge, including the knowledge of nature, to good Christians in heaven. But meanwhile, Boyle had to consider imperfect knowledge and imperfections in earthly apparatus. He had to control leakage of gases from lutes or from air pumps, control the different heats produced by furnaces, develop more accurate thermometers, and generally occupy himself with the operation and improvement of instruments.

Besides distinguishing between experiment and observation, Boyle divided experiment into two groups, "probatory" and "exploratory." Probatory experiments were designed not to test theoretical knowledge but to use such knowledge to test the reliability of the conditions surrounding an experiment. Those conditions included the purity or impurity of substances used and the adequacy of apparatus. Boyle made use of chemical indicators to test the progress of some reactions, so that, for example, he could decide whether a sample of spirit of salt (hydrochloric acid) was pure or contaminated.

Exploratory experiments could test hypotheses, see whether popular beliefs were well founded, or involve the invention of instruments, which could produce new phenomena by "reducing nature to alter her course." The same experimental procedures in different contexts could be either exploratory or probatory; distillation, for example, could determine whether a drug was pure or could be used to discover the drug's chemical constituents.

Not the least important of Boyle's practices concerning experiment was his habit of reporting experimental failures and disappointments. Failed experiments could suggest new lines of research or improvements in technique. Also valuable was his insistence on avoiding imprecise and arbitrary language in reporting the results of experiments or in framing hypotheses. Because Salt and Mercury simply had too many different meanings for Paracelsians, even when the experimental practice was good, the accounts of experiments and the reasoning from them were flawed.

Who Believes in Elements?

For Boyle, chemical observation and experiment could contribute significantly to natural philosophy by probing nature, exploring the inward parts of matter, and inquiring into causes. He pursued a program of "associating Chymical Experiments to Philosophical Notions."* He wanted to convince natural philosophers that chemistry could assist them, and part of his strategy for doing this was to show the congruence between chemical understanding and the mechanical philosophy. He wanted "to beget a good understanding betwixt the Chymists and the Mechanical Philosophers, who have hitherto been too little acquainted with one another's Learning."† He was, however, confronted with serious problems when it came to claiming that chemistry was a respectable part of the new natural philosophy of the seventeenth century. What were Boyle's colleagues in the recently founded Royal Society of London to make of proliferating and competing theories of the elements, textbooks that listed unreliable and unconfirmed observations, alchemical cheats and frauds, and the association of chemistry with secrecy, magic, unintelligibility, and downright dishonesty? Although there were a lot of chemist-alchemists in the Royal Society, such abuses would scarcely encourage philosophers to take chemistry seriously. Boyle had to establish a clear distinction between chemistry pursued according to the canons of his experimental philosophy, and the wrong-headed kinds of chemistry that had given the subject a bad name.

In 1661 Boyle published *The Sceptical Chymist,* which demolished what he regarded as either fallacious reasoning or incompetent experiment or both. It was above all an attack on theories of the elements devised by those seduced more by theory than by experimental evidence. We have already encountered the principal element theories that Boyle attacked. Aristotelians had their four elements, earth, air, fire, and water. Paracelsians had three, the *tria prima* of Mercury, Sulfur, and Salt, which were not the same as the common mercury, sulfur, and salt of the laboratory, apothecary's shop, or even (in the case of table salt) the kitchen. Van Helmont had either one or two, depending on how you interpreted him: water, or, taking the biblical account literally, water and air,

*Boyle quoted in Lawrence M. Principe, *The Aspiring Adept: Robert Boyle and His Alchemical Quest* (Princeton: Princeton University Press, 1998), 181.
†Ibid., 183.

which came before earth and were therefore in a sense primary. But then one had to remember the confusing circumstance that water, even if an original substance and so after a fashion elementary, nevertheless contained its own mercury and sulfur; and air was not regarded as a chemical species at all, although it might contain chemical substances in, for example, particles of smoke. The preceding sentences only make sense if the reader knows what meaning to attach to the words *element, mercury, sulfur,* and the rest. Using the same words with a variety of meanings was something that had offended Bacon and similarly offended Boyle, at least in those areas that he believed should be public science. There were regions of chemistry that he thought were best reserved for adepts and where secrecy was therefore acceptable, even desirable. But Boyle's *Sceptical Chymist* deals with public science, where theories are on trial, experiments have to be repeatable, and evidence needs to be confirmed by reliable witnesses.

Boyle took the various element theories literally, subjecting them to rational criticism and comparing their predictions with his own experiments and observations and with those of others. Heat was one of the keys to chemistry. It was common knowledge that heat was produced in most chemical reactions. A variety of fuels and furnaces had been developed over the centuries, giving the chemist control over the temperatures of *his* reactants and over the length of time they were heated.* In 1556 Agricola had written an encyclopedic treatise on metals which contained a great deal of information about the use and construction of furnaces. Such information, along with what we know about the traditional practices of alchemy, meant that chemists and alchemists knew how to use fire. They used fire to make alloys, to make charcoal, and in almost every operation of chemical and alchemical art.

We have seen that one of Boyle's major concerns was to reveal the inner workings of nature and that he considered chemical analysis to be one way to pursue this goal. Fire provided the commonest means of analysis, and Boyle was determined to show that, when applied to the various element theories of his day, it failed singularly to provide support for any of them. If one burned wood, ash and soot were produced, and so was smoke. This was not helpful

---

*I have italicized the word *his* to make the point that the early Fellows of the Royal Society, as well as physicians, surgeons, metallurgists, and mining chemists, makers of gunpowder, and even, in the main, apothecaries and alchemists were almost all men. There are a few exceptions, like Maria the Jewess, an alchemist in Alexandrian times, and Margaret Cavendish, Duchess of Newcastle, a corpuscular philosopher of the seventeenth century who is often referred to as Mad Meg. There are more women associated with chemistry in the eighteenth and nineteenth centuries. But it was not until the nineteenth century changed into the twentieth that women began to practice chemistry in significant numbers.

Renaissance Metalworking

Refining metals was important for alchemy but also for industry. The most famous manual of Renaissance metallurgy was by Georgius Agricola (1494–1555), who wrote a detailed account of mining technology. In the tenth chapter of this work, he explained how precious metals were separated from base metals and from one another. The techniques that he discussed often involved the use of acids, as in the separation of silver from gold. They also involved furnaces.

In this engraving, *A* is the furnace; note that there are several furnaces, since heating was the principal tool for metallurgists, just as it was for alchemists. *E* is the draught hole underneath the furnace, *C* indicates the air holes, and *B* the round hole in which a crucible (*F*) or distillation apparatus (*G* and *H*) could be placed. *K* marks the flasks in which the distilled substance could be condensed and collected.

There were many types of furnaces and of distillation apparatus, but every piece of apparatus shown in this sixteenth-century illustration could have been found in earlier medieval laboratories or in laboratories as late as the eighteenth century. Alchemical, chemical, and metallurgical laboratories were relatively unchanged for hundreds of years.

■ From Georgius Agricola, *De re metallica* (Basel, 1556), book 10, figure 1.

---

for any of the element theories. If one subjected recently cut green logs or branches of wood to destructive distillation, then things looked more promising. With distillation, one of the products was charcoal, which being solid might correspond to earth. Various liquid fractions were produced and could be condensed. Boyle characterized these as oil, vinegar, and water, which might correspond to the element water. Something that was hard to contain, which Boyle called spirit, also emerged during the distillation, and this might be related to air or water. When the distillation was carried out at a hot-enough temperature, the wood or charcoal would catch fire. Here perhaps was the element of fire. So it was just possible to claim that four elements were released from wood by fire, as long as one carried out the analysis under the right conditions. Merely burning the wood would not reveal so many elements. ⟩

There was a problem, though, which Boyle pointed out, taking his objection from Van Helmont: How do we know that what is produced by our analysis was previously present in the substance being analyzed? Analysis on its own does not tell us what substances preexisted in the compound substance being analyzed. But if we do not worry about that problem (and we *should* worry about it), maybe we can claim that the destructive analysis of wood shows that wood is composed of four elements. Boyle would not allow this. Fire analysis, he insisted, when applied to wood gives us not elements but "mixed bodies, disguised into other shapes: the *Flame* seems to be but the sulfurous part of the body kindled; the *water* boyling out at the ends is far from being elementary water, holding much of the salt and vertu of the concrete. . . . The *smoake* is so far from being *aire,* that it is as yet a very mixt body, by distillation yielding an oile, which leaves an earthe behind it; that it abounds in salt, may appear by its aptness to fertilise land, and by its bitterness, and by its making the eyes water."*

Fire analysis falls apart completely in the case of gold; no matter how much we heat it, we are left with the same metal that we started with. To make things worse, if we heat gold and silver together, we get an intimate mixture of the two metals, a kind of alloy. We can separate them again by using *aqua fortis* (concentrated nitric acid). Fire analysis gives us absolutely no basis to argue that gold is composed of four, or three, or for that matter of any other number of elements greater than one. Blood, in contrast, appears to give five products of analysis: phlegm, spirit, oil, salt, and earth. Are these five substances elementary? By means of such arguments, Boyle arrived at the conclusion that we have no good reason to adopt any element theory that claims there are universal elements present in all bodies. Fire analysis yields different answers for different substances and for the same substance under different conditions. So much for Aristotle and Paracelsus.

Boyle is, however, more impressed by Van Helmont, whom he regarded as a good experimentalist and whose results were generally reproducible. He was inclined to look favorably on Van Helmont for a variety of reasons, including his biblical literalism and his publication of the results of weighing reactants and products. Boyle took Van Helmont's willow tree experiment (see Chapter 1) a step further. If water was the source of the growth of the sapling and of the resulting increase in its weight, then there was no need to grow the plant in earth. So Boyle grew seedlings in water and confirmed that they increased in weight. He would not, however, accept water as the universal element. Since Van Helmont had specified that "elementary" water contained its own sulfur and mercury, he was not arguing for water as an element in the purest sense. In any case, Van Helmont's theory, like all the other element theories that Boyle

*Boyle quoted in W. H. Brock, *The Norton History of Chemistry* (New York: Norton, 1993), 57 (emphasis in original).

criticized, could not be demonstrated universally. Water was therefore no more a universal element than the *tria prima*.

One other important point about Boyle's attack on element theory is that he did think it was important for chemists to pursue their analyses as far as possible, so as to break the substances down into the simplest constituents that could be reached in the laboratory. A century later, Boyle's simplest products of analysis were to become the building blocks of a new system of chemistry.

Wonderful Atoms

Corpuscles might well be divisible, but atoms by definition were not. Today, we associate atoms with chemical elements. Boyle definitely did not do so. That made for difficulties in relating theory to laboratory practice, for reasons we shall soon see.

There were several versions of corpuscular philosophy in Boyle's day. Boyle focused on the shared aspects of different corpuscular and mechanical systems, including that of Descartes, and argued for their strengths when compared with Aristotelian and other nonmechanical doctrines. It is therefore not necessary here to explore the variety of mechanical and corpuscular systems available in the mid-seventeenth century, but it will be useful to note the principal ones. Gassendi had produced a version of Epicurus that was increasingly acceptable to Christians. Descartes had come up with the remarkable notion that spatial extension was equivalent to matter, so there could be no vacuum. There were other versions of atomism that left no room for God or for spirit, but such atheistic views were very unpopular. Isaac Newton had his own version of atomism, in which atoms coexisted with spirit and were ruled by God's agents; that is where gravitation and other forces came in, and they operated throughout all space. Boyle's corpuscular theory was, like Gassendi's and Newton's different brands of atomism, compatible with God and spirits.

As an older contemporary of Newton, Boyle did not have the advantage of Newton's speculations that chemistry might be handled by a system of short-range forces operating on atoms. He believed in atoms moving in space. Some of his most widely reported experiments were those carried out with an air pump. He explained that it was possible in the jar of the pump to produce a vacuum, where that meant simply a space altogether or at least almost entirely devoid of air. Boyle, unlike Descartes, believed in the possibility of a vacuum or void, a space empty of matter. He thought that all material bodies were made up corpuscles, which were in turn made up of atoms combined in different ways. It was from these different combinations of atoms that the qualities of bodies arose.

Gassendi adopted the medieval idea of "seminal virtues" which fitted atoms together and shaped them uniformly. Boyle carried out a very large number of experiments, on acids, alkalis, metals, crystals, and other substances, and concluded that each specifically different substance had its own particular internal form or virtue. That accounted for the uniformity of properties or qualities associated with and indeed essential to each distinct substance. The smallest atoms formed aggregates, which came together to form more complex aggregates. All the substances handled by an experimental chemist were compounded of simpler atoms, and their distinct properties arose from four properties of the corpuscles, their *bulk, texture, shape,* and *motion.* Bulk, shape, and motion are straightforward for us. We can understand an explanation of the sweetness of sugars in terms of the roundness of their particles and of the sharp taste and corrosive nature of acids by the geometrical sharpness of their constituent particles. Texture is a little more complicated. The word has the same root as the word *textile* and means the weave of bodies. The texture of particles helps to determine the way that they are woven together into bodies within the reach of our senses. Velcro fasteners are an obvious twentieth-century instance of texture or structure affecting the way bodies fit and stay together.

Boyle now had the foundations for his explanations of chemical change and material transformation. When a fertilized hen's egg produces a chick, there has been a transformation and reorganization of the egg-stuff into chick-stuff. The transformation is produced by a seed, and it is mechanical (rearrangement) and vital. Mechanical chemistry and a kind of vital chemistry, a chemistry of life, are both at work. Again, if a farmer with an orchard grafts a pear shoot onto a plum tree and the graft takes, then the same tree will produce both pears and plums. Here, clearly, the same substance is feeding the growth of both fruits on the same tree, and an explanation in terms of corpuscular rearrangement is one that Boyle found satisfying. Mechanical explanations, he argued, are simply so much more satisfying than explanations based on Aristotle's philosophy. They can account for anything, at least in principle. That is their strength—and also their weakness, as we can see, even if Boyle could not. An explanation that claims to cover everything really explains nothing, since it cannot be tested or refuted. Boyle had no way to prove the correctness of a particular mechanical or corpuscular explanation. He just had a conviction that this was the right *kind* of explanation. He could classify substances according to their qualities, as he did using indicators to discriminate among acids, base or alkaline substances, and neutral substances. But he could not use corpuscular explanations to make firm predictions, and he could not relate his laboratory classifications to any definite account of constituent atoms. If a theory has no predictive value and no firm correlation with practically derived classifications of substances, then it cannot take us very far.

Boyle's contributions to chemistry were numerous and significant. He advanced chemical classification a long way, and his category of neutral substances was valuable for an understanding of the chemistry of salts. Making chemistry a respectable part of natural philosophy was of great importance. The experimental method that Boyle devised, with its emphasis on evidence, repeatability, public verification, quantification, and the use of pure materials, was of even greater importance. His mechanical explanations, however, were ultimately sterile.

3 A German Story

What Burns, and How

Robert Boyle and Isaac Newton wanted to explain the properties of substances in terms of the properties of constituent corpuscles, which might be complex aggregates or else simple atoms. There was no contradiction between this approach and the alchemical ideas that they built into their chemical practice. But although they both advocated mechanical explanations in chemistry, they meant different things by *mechanical.*

We have seen that Boyle's mechanical explanations were ultimately sterile, although they had their supporters, as did other versions of corpuscular chemistry. These supporters, especially in France and England, argued, as Boyle had done, for providing what were essentially physical explanations for chemical phenomena, based on the shape, size, motion, and arrangement of atoms and groups of atoms.

Newton's explanations were, in contrast, based on the idea that bodies acted and interacted through their forces or active principles. He had stated, in the preface to his greatest work, *The Principia: Mathematical Principles of Natural Philosophy,* that this was the way to proceed: "For the basic problem of philosophy seems to be to discover the forces of nature from the phenomena of motions and then to demonstrate the other phenomena from these motions. . . . If only we could derive the other phenomena of nature from mechanical principles by the same kind of reasoning!"* The *Principia* had shown that gravity was one such force. It might well be that the cause of fermentation was another. In chemical reactions, such as fermentation, forces operated at very small distances. An understanding of the operation of these short-range forces would provide an explanation for chemical phenomena.

Some of Newton's younger colleagues claimed airily that it would not be a major problem to find the exact mathematical form of chemical laws. It was tempting for eighteenth-century chemists to find encouragement in Newton's personal authority, in the dramatic success of his work on gravity, and in the

*Isaac Newton, *The Principia: Mathematical Principles of Natural Philosophy,* trans. I. Bernard Cohen and Anne Whitman (Berkeley and Los Angeles: University of California Press, 1999), 382.

attractive notion that the uniformity of nature meant that chemical laws would
be like physical ones. A study of chemical reactions would, or so some believed,
lead to the discovery of the laws of force that determined the course of those
reactions. Just as an understanding of gravitational attraction made it possible
to predict the motions of bodies, so an understanding of attraction in chemistry would make it possible to predict the outcome of chemical reactions

In England, and to a lesser extent on the Continent, lectures were read and
books were written promising success in building a Newtonian chemistry.
These promises proved false, but that did not stop natural philosophers from
trying. Many seventeenth-century natural philosophers had thought that all
valid scientific explanations needed to fit into mechanical or corpuscular philosophy, terms used to describe a variety of theories of matter. When the discipline of physics emerged from natural philosophy, there were many physicists
who still believed that their science would provide the only real explanations
for chemistry. The eminent astronomer and physicist Pierre Simon Laplace
(1749–1827), who worked in the late eighteenth and early nineteenth centuries,
achieved a splendid synthesis of Newtonian astronomy and wrote books such
as *Celestial Mechanics* and *The System of the World.* Newton had built a theory
of gravitation that explained the motions of planets and comets. Subsequent
improvements in astronomical observations, and the rigorous mathematical
analysis of these observations, showed that Newton's theory worked even better than it had seemed while Newton lived. Laplace's books, with titles echo-
ing Newton's claims and methods, were a triumphant vindication of Newtonian astronomy. For a while, Laplace thought he could construct a chemical
mechanics to match the celestial mechanics. He too found that he was wrong
about chemistry, and he eventually admitted it.

Mechanical chemistry, whether it followed Boyle, Descartes, or Newton,
was to prove a dead end. It was largely although not entirely confined to Britain
and France. But there were lots of chemists throughout Europe, even including Britain, who were not seduced by dreams of mechanical explanations.
Some sought to avoid simple categories and boundaries and argued for more
complex explanations, drawing from chemistry itself, but also from physical
natural philosophy and from medicine. For example, the extremely influential
Dutch professor Hermann Boerhaave (1668–1738) drew on more than one discipline in constructing his chemical explanations. Another group of chemical
theorists and practitioners regarded chemistry as an autonomous science, and
they took pains to make clear its independence as a discipline. They will appear as major players in Chapter 4.

Yet another important group, strongest in Germany but active throughout

Europe, thought that chemistry was part of medicine, or even that medicine was part of chemistry. That was the program of the medical chemists, or iatrochemists, who followed the lead first offered by Paracelsus and then modified and extended by Van Helmont. They carried on with their chemistry or iatrochemistry through the seventeenth and into the eighteenth century, and they took precious little account of what mechanists were doing.

Chemical Cures and Medical Disasters

Paracelsus revolutionized alchemy and medicine at a single stroke. He rejected the great physician Galen with his four humors; he rejected the great philosopher Aristotle with his four elements; and he rejected the theories of those who by the sixteenth century had become traditional alchemists, with their dyad of Sulfur and Mercury. The Renaissance, at its peak in Paracelsus's day, was a time when ancient learning was made new. It was also a time when ancient learning was displaced by truly new learning. Conservative and classical scholars made their contributions, as did revolutionaries and iconoclasts. Paracelsus was a revolutionary, and he gloried in portraying himself as one. Instead of humors and traditional elements, he advocated his *tria prima*, happily ignoring the fact that two of his three principles were old. He gave a new spin even to the old principles. Mercury was the volatile and watery or fluid principle, Sulfur was oily and inflammable, Salt was dry and the source of flavor. The new principle, Salt, suitably interpreted, was to be of great importance in the history of chemistry, especially in the eighteenth century (see Chapter 4).

Alchemy, in its new Paracelsian form, was once again to provide the key to medicine. The fundamental operation of alchemy, whether achieved by calcination, distillation, or otherwise, was that of separation, and that remained true of the chemistry that came after alchemy. (At the beginning of the twentieth century, *Scheidekunst,* meaning literally the art of separation, was a German word for analytical chemistry, and the Dutch word *scheikunde,* also derived from the words for separation and for art or skill, still means the whole science of chemistry today.) Alchemy could separate out impurities, converting impure to pure substances, separate beneficial from harmful substances, and turn useless or noxious substances into healthful medicines.

Paracelsus's notion of healthful medicines was sometimes alarming. He pioneered the use of mercury compounds in the treatment of syphilis, and he sometimes achieved spectacular cures. He also managed some spectacular kills of his patients, and more than once he had to leave town in a hurry. He concentrated on metals and minerals in the treatment of sickness, and so was able to develop a list of medicines, a pharmacopoeia, which was very different from the traditional Galenic ones. Galen with his four humors had sought general remedies that would adjust the humoral balance of a body that was diseased, disordered, and out of balance. His remedies were often herbal, which means organic, complex, and chemically very hard to describe. Paracelsus's remedies, in contrast, were often mineral or inorganic, simple in character and composition, and therefore chemically relatively easy to describe. Those physicians who followed Paracelsus's approach were known as iatrochemists. They held alchemy in high repute and were keen to learn about and to contribute to the chemistry of minerals and metals. They also studied mineral waters, and the later vogue for health spas in Germany and elsewhere may be traced to their influence.

Paracelsian physicians were likely to find themselves in opposition to the established medical fraternity. Medical doctors in continental Europe by the early seventeenth century had to choose between Galen and Paracelsus, and predictable institutional and professional conflicts, rivalries, and splits were the result. English Paracelsians managed to compromise, performing the improbable trick of grafting Paracelsus's ideas about chemical medicines onto Galen's humoral pathology. But the English Paracelsians were not the ones who made the most significant contributions to chemistry. Continental Europe, especially the German-speaking parts, was where the action took place from the late seventeenth century to the mid-eighteenth century.

Iatrochemistry, Alchemy, or Chemistry? Johann Joachim Becher

It is obvious when we read chemical texts that by the mid-eighteenth century, alchemy has, for almost everyone, been left behind. That process began in the seventeenth century, although two of the greatest English natural philosophers of that century, Boyle and Newton, devoted much effort to the study and pursuit of alchemy. Johann Joachim Becher (1635–82) was a near contemporary of Boyle, born a few years later and dying a few years earlier than the Irish-born chemist. Unlike Boyle, he did not inherit great wealth and therefore had to make his own way in the world. He did so by a combination of business acumen, chemistry and alchemy, and medical practice. An important step was his marriage to the daughter of an imperial councilor, which transformed his career prospects. He was for some years the court physician at Mainz in the Holy Roman Empire, and then became an imperial commercial councilor in Vienna. He soon had responsibilities that were far more than medical—he had to negotiate agreements and investments for the emperor, he organized a major eastern trading company, and he built and oversaw glassworks and textile factories. He was a very accomplished polymath.

Before we look at Becher's chemistry, it is useful to think for a moment about the assumptions involved in writing and reading the history of chemistry. The history of any science is a story of progress. Chemists today understand the behavior of chemical substances in their reactions better than any of the great chemists of previous centuries did, just as Boyle knew more chemistry and more chemical substances than the greatest Alexandrian alchemist. We study the history of chemistry knowing that modern chemistry is better than old chemistry, and that old chemistry, like any old science, is frequently wrong. That, as we have already seen, does not mean that it was not good science in its day. Newton's achievement is not lessened because Einstein's theories of relativity represent an advance over Newtonian mechanics. But we cannot wholly ignore what modern scientific knowledge tells us about old science, nor should we do so. It is tempting, then, to concentrate on those parts of old science that we can see as leading to later science and eventually to our modern science. We have to do this to some extent in order to recognize progress in scientific knowledge. But Boyle and Becher did not know where their chemistry or alchemy or chymistry would lead, and so they valued their discoveries in ways different than we might. Somehow we have to try to combine our understanding of progress with knowledge of what was important to earlier scientists. That is why it is necessary to look at Boyle's alchemy alongside his mechanism and ideas about scientific method, and together with his practical and conceptual contributions to later chemistry. Boyle is not modern, although he has often been presented as if he were.

Neither is Becher modern, but he has, until very recently, been mainly presented as one opposed to intellectual enlightenment, an alchemist whose unintelligible writings somehow inspired later chemists to develop more modern and intelligible theories. We need to recognize that he was an alchemist and an iatrochemist and that he was at the same time seeking practical solutions to what we can recognize as legitimate chemical problems. That often meant making use of his alchemical knowledge and the theories that underlay it. When it came to the scientific resolution of medical issues, his adherence to iatrochemistry, or medical chemistry, was of central significance.

Becher's ideas were in a line that Paracelsus had initiated, but they were not ones that Paracelsus would have recognized. As had Paracelsus, Becher believed that chemistry was the key to medicine. He also believed that there were three elementary principles, but these were not Paracelsus's *tria prima*. Instead, Becher opted for air, earth, and water. Air, for Becher as for most seventeenth-century chemists, was not truly a chemical principle; Becher regarded it as an agent for mixing chemical principles and the substances that were made up of them. So earth and water were really his principles or elements and, mixed with air, they generated organic bodies and minerals. Metals grew from seeds in the earth. The alchemical foundation is a familiar one. But then Becher added a refinement. He decided that three different kinds of earth were needed in metals and minerals. One kind of earth in particular accounted for combustibility. It was oily and inflammable, and he sometimes referred to it as inflammable sulfur, in a tradition that goes back to the medieval alchemists, and even to Aristotle. He, as Van Helmont had done before him, used a Greek word, *phlogistos,* for inflammability. The idea of an inflammable sulfur as the cause of combustibility in substances was to prove important in eighteenth-century chemistry. Under the name "phlogiston," it was to outlive Paracelsus's *tria prima* and Becher's other kinds of earth.

Georg Ernst Stahl: A Chemist of Principles

The chemist whose name is most closely associated with phlogiston was Georg Ernst Stahl (1660–1734). He was—like Paracelsus before him, and like Becher, whom he much admired—both a physician and a chemist. He was not only employed as physician to the king of Prussia, where, in Berlin, he was the first royal physician; he also taught as a professor of medicine at the University of Halle. He wrote textbooks for physicians. But in contrast to Paracelsus, he did not believe that chemistry was the key to medicine. He believed that living bodies had a soul that controlled the transformations of matter within the body, including chemical processes. The science of chemistry was fine for explaining the reactions of substances like minerals that had never been alive, and it could account for their composition and decomposition. Chemistry could also explain the way formerly living bodies rotted and decomposed after death, but it was no good for understanding their preservation and their material transformations while they lived. The soul, or vital principle, was what enabled living bodies to fight against the processes that led to decay.

By arguing in this fashion, Stahl advocated a form of *vitalism,* a doctrine that holds that living bodies are alive by virtue of a vital principle. Van Helmont had also had the notion of an active spirit in bodies which worked locally in the body to produce chemical change and affect health. The importance of the idea of local action was that it enabled physicians to break away from the general pathology of the Galenists and to consider specific cures for specific illnesses. Van Helmont described a different active spirit for each of many different parts of the body. In this respect Stahl's view of the local action of the soul can be seen as similar to Van Helmont's. But in his refusal to allow medicine to be explained by chemistry, Stahl broke from traditional iatrochemistry. Indeed, he had much in common with an emerging breed of chemists who wanted to claim independence for their science, free from physics or medicine.

Stahl was, however, precisely like Paracelsus in one respect: he enjoyed a good and vigorous argument. One of his targets was the application of mechanism to explain medicine or chemistry. He was against what philosophers today call reductionism, in which the phenomena of one science are explained in terms of the phenomena of another science that is somehow regarded as more basic or fundamental. So, as far as he was concerned, chemistry and mechanics were unrelated. It thus followed that all kinds of mechanical chemistry were mistaken, and they were all as bad as one another. Corpuscular philosophers like Boyle, who believed that matter and motion explained chemistry, and dynamical philosophers, who followed Newton in explaining chemistry in terms of the interaction of matter and *forces,* were equally wrong in believing that their natural philosophy could account for chemistry and medicine. Stahl wanted a truly chemical philosophy, a philosophy of nature based upon chemistry; and he wanted to make sure that it was kept uncontaminated by either medicine or physics.

The key to the difference between chemistry and physics lay in the distinction Stahl made between aggregates or mixtures and what he called *mixts.* Becher had made this distinction, but he had not elaborated it as carefully and explicitly. Mixts for Stahl were true chemical compounds. He saw the main business of chemists as the analysis of mixts and the characterization of their constituents. A chemical compound, or Stahlian mixt, was one whose properties were not simply the sum of the properties of the substances that had come together to form it. Their formation involved changes in the chemical properties of constituents that are modified in combination, as when an acid and a base or alkali react to form a neutral salt. Acid and base both change their properties when entering into a mix or compound, and most salts have neither acidic nor basic properties. Aggregates or mixtures, in contrast, come into existence by the mere physical mixing of unchanged constituents. Stirring two different kinds of sand together would produce aggregation; the different kinds of grain could be separated physically, for example, using tweezers, without resort to chemical instruments. The properties of this physical aggregate were simply the sum of the properties of the two kinds of grains of sand. Chemical change was brought about by *mixtion,* not through the action of physical forces such as gravitation or cohesion.

Saying that chemistry cannot be explained by physics is important, but it takes more than a negative statement to construct a chemical philosophy. Stahl set about providing the missing ingredients. Like those of any innovator, they owe much to his own invention but also much to his predecessors. Stahl drew principally from German chemical traditions and pretty well ignored what was happening in France (see Chapter 4). It was Becher who influenced him most.

Stahl's view of the cause of chemical reactions had a very respectable antiquity. He relied on the doctrine of affinities, which traditionally meant the chemical attraction of like for like. It was only later that affinity came to mean the chemical attraction of opposites, for example, of acids for alkalis or, in the nineteenth century, of electronegative for electropositive substances (see Chapter 7). Metals could form alloys, often regarded by chemists as mixts or compounds, by virtue of their shared metallic character. Metals could be dissolved in acids because acids and metals had something in common, some shared principle. That was far from apparent, and it was up to chemists to demonstrate that it was true. Chemists would use the methods and tools of analysis to identify the constituents of mixts.

For Stahl, as for Becher, water and earth were the key principles of chemical substances. Like Becher, Stahl said that there were three different kinds of earth, but he described these somewhat differently from his mentor. There was metallic or mercurial earth, which accounted for the brightness and malleability of metals, their ability to be molded and worked by a goldsmith or blacksmith. Then came vitrifiable earth (earth that can be turned into a glassy substance), and this was what made substances able to melt. It was associated with the heavy, lumpish nature of minerals. Finally, there was sulfurous earth, also known as phlogistic earth or phlogiston, and this enabled bodies to burn and flame. These three earths were chemical principles.

What were "principles" for Stahl? They were not universal elements, like Aristotle's four elements or Paracelsus's *tria prima*. They were, however, like Aristotle's elements in one crucial respect: although they were material, they could not be isolated. Most importantly for Stahl, they were the causes of particular properties of chemical bodies, and they conferred those properties on the mixt bodies that contained them as constituents. Bodies burned if they contained the phlogistic earth or principle. If they did not contain that principle, then they could not burn. The phlogistic earth could therefore reasonably be called the principle of combustibility.

So far, this does not go a long way beyond Becher. But it was Stahl's special contribution to bring together two classes of chemical phenomena that had previously been viewed as quite different. Any scientific explanation that is capable of explaining more kinds of phenomena than its predecessors and that does not contradict any known evidence has to be considered an advance.

Stahl interpreted the rusting or corrosion of metals as a kind of slow combustion, like the burning of wood or charcoal in a fire or furnace. In both cases, he argued, the burning bodies lost some of the phlogistic earth that they contained. That, indeed, was essentially a definition of combustion, a process involving the loss of phlogistic earth or phlogiston.

If corrosion was a process involving the loss of phlogiston, then the transformation of ores to metals, as carried out by metalworkers, had to involve the restoration of phlogiston. Charcoal, readily combustible and therefore rich in phlogiston, could be smelted with metallic ores low in phlogiston. By giving up its phlogiston to the ores, the charcoal transformed them into metals. Here we have a view of metals as compounds or mixts (containing at least phlogiston and mercurial earth) and of their ores as *simpler* substances. One of the simplest kinds of metal ore is identical with the substance produced when a metal is heated. We call it an oxide; eighteenth-century chemists called it a calx. Heating the calx with phlogiston-rich charcoal could reverse the transformation of metal to calx.

These transformations of metals and their ores or calxes were not all that Stahl's notion of principles explained. The combustion of sulfur yielded an acid, which Stahl, following Becher, called the universal acid, because he considered it to be the *principle* of acidity, the material constituent that was essential to the formation of every acid. Since sulfur could burn, it must contain phlogiston. Today, we consider sulfur, like the metals, to be a chemical element. For Stahl, the universal acid was a mixt. So were sulfur and the metals, because he believed that they contained more than one material constituent. That is, they were more complex than the products of their combustion.

Modern chemistry tells us that this is the wrong way around. When sulfur burns or when a metal corrodes, the substance produced is more complex than sulfur or metal. But a wrong theory, as we have already seen, can still be useful. This one is indeed useful because it explains a wide range of phenomena. And its explanatory range is not limited to what we have just considered. It can answer the question, Why does a candle or other flame go out after burning for a time in an enclosed space? In order to answer this question, we need to understand how Stahl regarded atmospheric air. He viewed it as Becher had done, not as a chemical species but rather as a physical environment in which chemical phenomena may take place. It operated like a sponge that can soak up water but only so much. A given volume of air can soak up only so much phlogiston. When it will not absorb any more, the fire or flame goes out. (The fire could also go out because the burning body has released all its phlogiston.)

One could imagine that with all the combustion that has taken place in the world, our air or atmosphere would become saturated and nothing could ever burn again. That has not yet happened. Why hasn't it? Stahl's answer was that plants absorb phlogiston from the air. So Van Helmont's willow tree grew not only because it absorbed and transformed the water with which he nourished it; the tree also grew by absorbing phlogiston from the air. That is why wood, and the charcoal that comes from it, can burn; it is rich in phlogiston. This is a long way from modern explanations based on the idea of photosynthesis, which chemists began to work out in the late eighteenth century. But it is a fruitful extension of an explanation, and so it makes phlogiston even more useful. In fact, Stahl used the concept of phlogiston to handle far more than combustion. The phlogiston theory, in Stahl's work and as it developed through the eighteenth century, became one of the most powerful and fruitful theories in the history of chemistry.

Besides principles and mixts, Stahl made much use of the idea of chemical instruments, by which he meant those mechanical agents that made mixts possible but were not their material cause, that is, were not the stuff of which mixts were made. Stahl's instruments included fire or heat, which was a necessary cause of so many chemical reactions. The controlled use of sources of heat had been one of the key skills for alchemists and chemists since their disciplines first took shape. Note that heat here is an instrument and not an element, as it had been for Aristotelians and others. Water could operate as a solvent, without entering into a mixt, and in that case it too could function as a mechanical agent, one of Stahl's instruments. Air, for example, when it absorbed phlogiston, could similarly be an instrument. Chemists used instruments as tools to produce or analyze mixts.

The idea of phlogiston, and the theory of combustion based upon it, originated in Germany. German chemists of the early eighteenth century were for the most part working from the Paracelsian tradition, even where, like Stahl, they transformed and transcended it. In their opposition to mechanical chemistry and in their efforts to develop a chemical philosophy, they not only rejected mechanical chemistry as it was shaped in England and France but also rejected attempts to purge chemistry of metaphysically rooted ideas such as that of affinity. They wanted to stick to their own chemical philosophy and were unhappy with the alternative philosophy of empiricism, which, based upon the work of Boyle and Newton, insisted on experience as the source of *all* knowledge. John Locke in England was the leading spokesman for this new philosophy. Empiricists claimed to have no time for things that could not be seen, heard, smelled, tasted, or felt. In the next chapter, we shall see how empiricism was central to the ideology of eighteenth-century French chemists.

But it is necessary to meet one of these chemists, Rouelle, here, because of the way he regarded principles and instruments. We shall encounter him again later.

Guillaume François Rouelle: Elements, Principles, and Instruments

Almost every important French chemist in the middle years of the eighteenth century attended lectures that Guillaume François Rouelle (1703–70) gave at the King's Garden (the Jardin du roi) in Paris. He was a lively lecturer, dismissive of the theoretical excesses of other lecturers, and anxious to make his lectures be *practical* demonstrations of chemical phenomena. His style was scarcely that of traditional academics. In the heat of his experiments, he would roll up his sleeves, get his hands and forearms and sometimes his face and shirt dirty, and show how chemistry was above all a science of practice. This was not just rhetoric; for Rouelle, practice was a crucial part of chemistry. Rouelle had more than important theoretical ideas to communicate. He had innovative views about principles and instruments, and these views brought together concepts that Stahl had kept distinct.

For Stahl, chemical principles and instruments were different. Principles, such as phlogiston, entered into the composition of mixt bodies and conferred properties on them. Instruments were mechanical agents that made mixts possible, but they were not constituents of mixts. Rouelle observed that phlogiston was associated with fire, the most apparent result of combustion. Fire was an instrument, but it was inseparably associated with phlogiston. Rouelle's principles functioned as instruments in chemical operations. They were also substances, which, like Stahl's principles, could not be isolated from the mixts that contained them.

The conception of phlogiston's dual role, as principle and as the matter of fire, was an extremely important result of Rouelle's modification of Stahl's ideas. It was Rouelle's version, and not Stahl's, that was to preoccupy chemists in the middle and final decades of the eighteenth century. We shall return to this version more than once in the following chapters. But first, we need to move from the German states of Europe to France and to the wider context of Rouelle's chemistry.

4 An Enlightened Discipline

Chemistry as Science and Craft

In 1726, Voltaire, an independent spirit who through his lively wit came to embody the French Enlightenment, ran afoul once again of a nobleman whom he had satirized in his writings. The nobleman's response was to have Voltaire beaten up. Voltaire challenged him to a duel, found himself (not for the first time) thrown into the prison of the Bastille fortress, and was released only when he promised to go immediately to England. There he immersed himself in literary life. He studied the philosophy of John Locke and, as far as he could master it, the physics of Isaac Newton. He decided that Locke and Newton, for whom the laws of nature were based on experience, had shown the right way to do science. Memories of his time locked up in the Bastille helped Voltaire to develop an enthusiasm for things and ideas English, to the disadvantage of things and ideas French. He also developed a taste for science that was to lead him, in later years, to install a chemical laboratory in his chateau, where he and his mistress explored the latest developments in chemistry.

In spite of Voltaire's experience, and in spite of the undoubted importance of the work of Isaac Newton, it was France and not England that became the cultural hub of eighteenth-century Europe. This was to be as true in chemistry as it was in most areas of thought and practice. The French took what they wanted from Locke and Newton and combined these ideas with the rationality and organization that epitomized their culture. The resulting intellectual climate was one that Voltaire increasingly represented. It was appropriately called the Enlightenment, because it represented the conscious rejection of authority, superstition, and magic in the light of experience and reason. Progress was to be the watchword—progress in knowledge and society, including its material aspects. Science and its applications, founded on experience, would undergo improvement and in turn would lead to the betterment of the material and moral lot of humanity. Chemistry was to play an important part in this process and progress. It was about to undergo one of its repeated transformations, to expand greatly, and to become, not for the first time, part of the dominant scientific culture of the day.

Voltaire was in many ways a radical, but even the French establishment under the old regime nurtured enlightenment and science. The Royal Academy of Sciences had been founded in Paris in 1666, soon after the Royal Society of London. Chemistry had a role in the Academy from its foundation; and chemists who were members of the Academy at the start of the eighteenth century were expected to contribute to the advancement of their science, assessing and improving its practical applications. There were also two professorships in chemistry, established in the seventeenth century, in the Jardin du roi in Paris. The lectures given by the professors were increasingly well attended by all who had an interest in that science, including philosophical chemists, metallurgists, dyers, apothecaries, physicians, and even some geologists. Chemistry was fully ready to perform as an enlightened science.

The Enlightenment was more than the Academy and the Jardin du roi. If Voltaire was its embodiment in person, then the great *Encyclopedia* of mid-century was its embodiment in print. If one talks about "the" encyclopedia today, different people will think of different encyclopedias. There was no such confusion in mid-eighteenth-century France. Everyone knew that "the" encyclopedia was the work edited by Diderot and d'Alembert, the *Encyclopedia, or Analytical Dictionary of the Sciences, Arts, and Trades.** Gabriel François Venel (1723–75), a pupil of Rouelle, wrote the article on chemistry. He told his readers that it was a mistake to seek to reduce chemistry to physics. Chemists had their own independent science, which could penetrate beneath the surface of things and get to their true nature, their inner essence. Physicists, in contrast, dealt only with external and accidental characteristics of bodies. Chemistry was and had to be an autonomous science, practiced by specialists.

Venel defined chemistry as "a science concerned with the separations and combinations of the constituent principles of bodies, whether effected by nature or by artifice, with the goal of discovering the properties of these bodies, or to render them suitable for different uses."† This definition still uses the language of principles that we encountered with Stahl and then with Rouelle. Clearly, however, with its talk of separation and combination, it refers to a science of operations, one in which the composition of bodies is seen as related to their properties. Bodies, indeed, are to be defined by their *constitution*, by the *reactions* that lead to their *composition* (or synthesis) and *decomposition*, and

---

*The French phrase translated here as "analytical dictionary" is *dictionnaire raisonné,* but the meaning of *raisonné* cannot be translated literally as "reasoned"; here it means based on critical, rational, analytical principles.

†"Chimie," by G. F. Venel, in *Encyclopédie ou dictionnaire raisonné des arts, des sciences, et des métiers,* ed. Denis Diderot and Jean le Rond d'Alembert, 28 vols. (Paris and Neuchâtel, 1751–72), 3: 408.

Chemistry in the *Encylopédie*

The *Encylopédie* of Diderot and d'Alembert was the manifesto of the French Enlightenment. It was an ideological statement as well as a wonderfully optimistic account of the role of science and its practitioners in the progress of civilization. *Progress* is one of the key terms in the Enlightenment, and it assumes the perfectibility of humankind and the improvement of material culture. It was at least the incubator of the ideas of democracy and equality, the intellectual precursor of the constructive side of the French Revolution of 1789.

Artisans and craftsmen would, through their empirical practice, reveal to gentlemen-scholars some of the workings of nature, and gentlemen-scholars would, through their theoretical or philosophical understanding, be able to help craftsmen to improve the operations of their trades. This exchange of expertise and insight is shown in the engraving depicting a chemical laboratory. Coats, cravats, and wigs identify the gentlemen-scholars in the laboratory just as aprons identify the artisans.

Note that the apparatus in their mid-eighteenth-century laboratory would not be out of place in an alchemical laboratory of the Middle Ages; the same range of crucibles, furnaces, and distillation apparatus (the last on the long shelf above the laboratory bench) could be found in each.

■ Denis Diderot and Jean le Rond d'Alembert, eds., *Encyclopédie. Recueil des planches.* Seconde livraison, en deux parties (Paris, 1763): Seconde Partie, "Chimie," plate 1.

by their empirically determined *properties*. Here, in a short space, was a whole program for the theoretical and practical development of chemical science. Chemists were the ones who changed bodies or who combined them so as to make them useful for practical ends, including industry, agriculture, and pleasure. Theirs was an enormously valuable enterprise, vital for society and commerce.

The plates illustrating the *Encyclopedia* were an integral part of the work. They showed craftsmen and artisans at work, while philosophers or *savants* (the Enlightenment term for scientists) watched and studied what they did. Some plates portrayed industrial processes and buildings. Others represented the interior of chemical laboratories, with their furnaces, distillation apparatus, and the rest; and yet other plates were devoted to a detailed portrayal of the latest chemical and related apparatus, including the balance. A central message, conveyed by the plates as much as by the text, was that philosophers could learn from the experience of practical men and that practical men in turn could improve their practice by listening to what philosophers had to tell them. Historians generally locate the Scientific Revolution in the seventeenth century and date the Industrial Revolution toward the end of the eighteenth century. The plates illustrating chemistry in the *Encyclopedia* strongly suggest that the Enlightenment enterprise helped to build a bridge between these two revolutions. They also suggest that chemistry was an important part of that bridge.

Scotland was another country in which chemistry thrived and which enjoyed its own age of Enlightenment. Scotland had been in a formal union with England since the early eighteenth century, and an attempt to restore the Scottish-derived house of Stuart to rule both England and Scotland was crushed on the battlefield of Culloden in 1746. But Scotland, or at least its capital city of Edinburgh, had long been closer to European culture than London was. Scottish intellectuals visited France and spoke French; Scottish doctors, in the early eighteenth century, took advantage of Europe's better medical education and studied in the Netherlands, where Boerhaave gave the best chemical lectures of the day as part of the medical curriculum. Those Dutch-educated Scots then returned to Edinburgh and Glasgow, where they gave chemistry a real presence in the universities.

Scots of the next generation were able to study at home. They maintained and strengthened chemistry in the universities from the middle of the eighteenth century on, in the years of the Scottish Enlightenment. Philosophers, lawyers, economists, literary men, chemists, and natural philosophers all contributed to the intellectual ferment in Edinburgh, and, as in France, theory and practice combined. Ironworks and agriculture were just two of the practical areas where academic chemists contributed to the passion for "improvement," the Scottish term for material progress. The leading figure in Scottish chemistry in the second half of the eighteenth century was Dr. Joseph Black, who had written an M.D. thesis on a possible chemical cure for bladder stones. That thesis involved the study of a salt, an examination of the nature of heat and its role in a chemical reaction, the use of the balance as a tool for chemical analysis, and the identification and characterization of a gas as a chemical species. In almost every aspect of this work, Black contributed significantly to key problem areas in the rapidly developing science of chemistry. In the next chapter we shall return to Black and see why his work was important. First, however, we need to look at two central areas of chemical investigation in the eighteenth century: the elucidation of the chemistry of salts, important for pharmaceutical and mineral chemistry alike, and the development and application of a new concept of chemical affinity.

Salts of the Earth, and the Classification of Substances

Paracelsus had introduced Salt as a principle in his chemical classification. For Stahl, salt was not so much a principle as a category, and his laboratory skills enabled him greatly to expand knowledge about salts. In 1723 he published a book on salts, in which he argued that they were produced by a combination of earths, alkalis, or metals with water. The book was reprinted, and a second edition followed, as did a French translation. Stahl's theory of salts, and his experiments and observations on salts, were thus available to European chemists throughout the middle fifty years of the 1700s. Stahl had an influential view of the constitution or composition of salts. He and other chemists recognized that an acid was involved in the composition of each and every salt, and, as a chemical philosopher, he regarded this acid as the most important part. But he saw that salts were of commercial as well as philosophical interest, and in their commercial aspect, other parts also needed to be identified (e.g., the "metallic" part).

Stahl's work became known in France. Meanwhile, French chemists within the Academy worked separately and independently on the chemistry of salts. As the eighteenth century progressed, German and French understanding of this area of chemistry came steadily closer together. The French translation of Stahl's book on salts marks the effectiveness of that union. The leading student of salts in Paris at the opening of the eighteenth century was Wilhelm Homberg (1652–1715), a widely traveled chemist who made Paris and its Academy his home. He developed an elaborate classification of salts. Composition and experiment were the keys. He found the definition of any given salt to be threefold. It depended on (1) the properties of that salt that the chemist could detect through the senses of sight, smell, taste, and touch; (2) the laboratory operations that led to the preparation of the salt; and (3) the substances of which the salt was composed. The chemist in his laboratory could, for example, combine one of the mineral acids (hydrochloric, nitric, or sulfuric acid) with an alkaline earth, such as the one found in lime, to form a salt.

The implication of Homberg's approach, which became widely accepted, was that a salt, or any other chemical substance, was defined by three things: the substances that composed it, the operations by which it was prepared, and the totality of its empirical properties. The practical correlation of operations in the laboratory with chemical constitution was of great significance because it led to a new understanding of composition. This was more complex than the older ideas of elements and principles and more useful than they had been in demonstrating the differences between chemical substances. And it was the work of Homberg and his colleagues that was absorbed into Venel's definition of chemistry that we encountered in the *Encyclopedia.*

The definition of substances in terms of their properties was philosophically problematic, although the problem was far from being a new one in chemistry. Some salts were produced by a vigorous combination of acid and alkali— a union of substances of chemically opposite character, in violation of the old idea of affinity as the cause of the union between like substances. But the salts produced were neither acidic nor alkaline. Chemical indicators such as litmus, which turn red in the presence of acids and blue in the presence of alkalis, showed clearly that most salts composed of acid and alkali had the properties of neither parent. Corpuscular explanations, harking back to the seventeenth century, were brought forward. Homberg, for example, suggested that we could think of acids as having pointed corpuscles, like daggers, while alkalis were the sheaths. Combining the two would then be like sheathing a dagger, concealing its sharp point.

Chemists had long relied on a rich imagery of similes and metaphors. Homberg was working in a fine old tradition. And corpuscular explanations, however metaphorically they were intended, did at least offer a way of thinking about the preservation of chemical constituents while their properties were concealed in the properties of the compound. Corpuscular explanations also made it reasonable to envisage the perseverance of a chemical constituent through a series of reactions, so that the same substance was carried from one reaction to the next, and from one substance to another, without undergoing essential change, even though its properties could be masked. Such explanations reinforced the idea that chemical *composition,* embodied in chemical substances and revealed through chemical operations, was the key to chemical *classification.*

An immediate fruit of this new way of identifying different substances was a rapid increase in the number of known salts, and indeed of new substances of every sort. Different alkaline earths were identified, and chemists discovered that there were two distinct caustic alkalis (soda and potash). New acids, new metals, and new combinations between them threatened an information over-load. The only way to handle the rush of information about newly discovered or discerned substances was to devise schemes of classification that would enable chemists to find their way through the ever-expanding knowledge. How else could one bring order to the threatening chaos of new discoveries?

There were critics who thought that chemistry could never be more than a combination of laboratory practice, which they viewed as a kind of cookery, and classification, which they saw as the essence of natural history, including botany and zoology. Those same critics regarded natural history as unscientific, lacking the rigor of mathematical physics or astronomy. Chemistry was indeed far from Newtonian physics. We have already seen the failure of attempts to assimilate chemistry to Newton's program. But to dismiss chemistry for this reason was to adopt too narrow a definition of science and to undervalue the role of classification in the scientific enterprise as a whole. Natural historians have to classify what they observe or collect; so do chemists.

Chemistry was a laboratory science, a science of practice. This had always been so, and Enlightenment pride in laboratory practice merely put a seal of approval on an established fact. But mere empiricism had never been enough in science, had indeed never been possible in science. A major component of science is the organization of knowledge in ways that lead to its refinement and expansion. The organization of knowledge requires some scheme of classification and a language, or at least a set of terms and rules for using them, in order to make the classification fruitful and functional. Classification is essential to science. Eighteenth-century chemists knew that as well as anyone. Chemical operations and an acceptance of the importance of composition to the definition of any compound substance were two of the essential supports of their schemes of classification. A language that embodied these notions was another support (see Chapter 6). Finally, there was a newly reformed notion that gave order to the mass of experiments and identifications: chemical affinity.

Affinities: Classifying Substances and Reactions

The creation of tables of chemical affinities was an attempt to encapsulate all possible reactions between the constituents of chemical compounds. The goal was not only to provide a summary and key to known reactions but also to predict reactions that had not yet been observed. Tables of affinities thus had both a descriptive and a predictive role; they could be used as a shorthand for a description and classification of observed reactions, and they could function as instruments of discovery. It was also possible, although not necessary, to use affinity tables as a clue to the mechanism of chemical reactions. It was along such lines that Isaac Newton had urged natural philosophers to reason from observed phenomena to the forces that caused them, and then to the laws that governed those forces.

The first affinity table to be published was that of Etienne-François Geoffroy (1672–1731), who had joined the Academy in Paris in 1699 as a student of Homberg and soon became an associate member of the Academy. By the time he presented his table of affinities to the Academy in 1718, he had an international reputation and had been elected as a foreign member of the Royal Society of London. It is important to recognize that Geoffroy was careful to call his table one of relations (*rapports*), not of affinities. He rejected old ideas of affinity as the sympathy of like for like. He was also anxious not to be identified with the Newtonian camp, where affinities were interpreted as the result of chemical attractive forces. The danger of such an interpretation was made clear by Bernard le Bovier de Fontenelle (1657–1757), who had been permanent secretary of the Academy since 1697 and was described by Voltaire as the most universal mind of his age. Fontenelle, for many years the official book and article reviewer for the Academy and writer of the Academy's annual *History*, speculated about what might be the cause of Geoffroy's *rapports*. "It is here," he wrote, "that sympathies and attractions would be altogether relevant, if only they existed."*

Geoffroy carefully avoided Newtonian attraction in writing his paper. He began with an account of the selectivity of chemical reactions. Different bodies had certain relations that led them to combine readily with one another. These relations, he asserted, existed in different degrees and obeyed their own laws. Experiments showed that in a mixture of substances, one substance would always combine with another particular one, in preference to all others. Displacement reactions, where one substance drove another out of a compound and took its place, provided an insight into this selectivity.

If two substances had an affinity for a third substance, then the one with the higher *rapport* for that third substance would be the one to combine preferentially with it. The idea of classifying substances by the degree of their tendency to combine with one another was not new. Stahl had hit on it, and so in a different way had Newton. There were, however, two important novelties in Geoffroy's formulation, besides his avoidance of the language of Newtonian attraction and the language of Stahlian affinities between like substances. These were, first, the potential universality of the tables of *rapports,* and, second, the predictive power of these tables. Universality was a goal that the mak-

*Fontenelle, *Histoire de l'Académie Royale des Sciences* (Paris, 1724), 35–37.

ers of tables of affinities never achieved, but they believed that if they could make the tables complete and universal, then all possible reactions could be deduced from them. Some predictive power was readily available, even using incomplete tables. If one knew the initial conditions, and if the reactants and their constituents were ranked in a table of *rapports,* then one could predict the chemical outcome.

As Geoffroy wrote, chemists would find in his tables "an easy method of discovering what happens in several of their operations, even when these are difficult to disentangle." Chemists would also discover "what *must be* the result of the mixtures that they make from different mixt bodies." Here, without causal explanation, Geoffroy was offering an interpretative scheme for chemistry that would have all the force of the laws of physics. Place a substance C in a mixture (generally in solution, i.e., dissolved in a liquid, usually water) of compound $AB;$ if C has a higher *rapport* for B than A has, it will displace A from its union with B. At the end of the reaction, BC will be the resulting compound, and A will have been expelled from its combination with C. For example, the second column in Geoffroy's table ranked metals in order of their reactivity with the acid from sea salt (our hydrochloric acid). Tin was placed above copper, because it could displace copper from its combination with that acid.*

Geoffroy's table was important, but it did not have many successors in the first half of the eighteenth century. There was one in 1730 and another in 1749. Perhaps French reluctance to identify *rapports* with attractions lay behind this lukewarm response. The second half of the century, however, saw a resurgence of interest in affinity tables, stimulated by an extremely influential textbook of 1749, Pierre Joseph Macquer's *Elements of Theoretical Chemistry,* which devoted a whole chapter to affinities:

All the experiments which have been hitherto carried out, and those which are still being daily performed, concur in proving that between different bodies, whether principles or compounds, there is an agreement, relation, affinity or attraction (if you will have it so). This disposes certain bodies to unite with one another, while with others they are unable to contract any union. It is this effect, whatever be its cause, that will help us to give a reason for all the phenomena furnished by chemistry, and to tie them together.†

*Geoffroy, "Table des différents rapports observés en chimie entre les différentes substances," *Mémoires de l'Académie Royale des Sciences* (1718): 202–12.

†P. J. Macquer, *Elements of the Theory and Practice of Chemistry,* trans. A. Reid, 2 vols. (London, 1775), 1: 12.

Newtonian ideas about chemical combination made inroads in France in the second half of the century, and affinity tables proliferated. By 1778, Macquer (1718–84)had decided that there were no separate laws of chemical affinity and that the law of universal attraction would suffice to explain the whole of chemistry, if only we could learn about the shape of the particles of bodies. In the same year, the second edition of the *Encyclopaedia Britannica* asserted that all theories of affinity were conjectural, "neither is it a matter of any consequence to a chemist whether they are right or wrong."[*] Here was a recognition that the utility of a scientific theory need not depend upon its truth. Affinity tables were above all *useful,* in providing a summary of existing knowledge about chemical reactions as well as a tool for predicting new reactions.

Reactions and Operations: Closing Circles and Enveloping Nets

Tables arrange data in significant ways. The terms listed exist in a defined relation to one another. Affinity tables list substances, define them in relation to composition, and embody our knowledge of chemical reactions. They are like dictionaries and encyclopedias that present knowledge and embody the interrelationship of terms. A work of reference such as a simple dictionary or encyclopedia might define a violin as a kind of small cello, and a cello as a kind of big violin. That circularity is fine, as long as we know something about either one of those musical instruments before we consult the work of reference. Chemical substances are also defined in relation to one another. Acids react vigorously with alkalis, some metals dissolve readily in certain acids, while others do not. If a substance is defined in terms of its reactions with other substances, we have a situation only marginally more complicated than the case of the cello and the violin. A network of cross references shows the unity of a set of definitions. The coherence of the network is complete when its set of definitions forms a closed circle. In both cases, we need to bring external knowledge to bear on our reading of the definition. In chemistry, that means we need to understand the conditions in which reactions occur and the operations needed to bring them about.

As a result, French chemists in the Enlightenment developed a double classification—one in terms of affinities and reactions, the other in terms of the conditions of reaction and the operations that caused desired reactions to take place. The classification in terms of affinities was printed in books and papers. The classification in terms of operations and experimental conditions was less formally expressed, but it was equally important. At the level of greatest gen-

---

[*]*Encyclopaedia Britannica,* 2nd ed. (Edinburgh, 1778), 3: 1808.

erality were two questions: (1) Was the experiment to be performed in the wet or the dry way? (2) How should heat be applied to assist the reaction? When chemists wrote of the "wet" way, they meant a chemical reaction in solution or between liquid reactants; the "dry" way involved reactions produced by the mixture of dry reactants. Mixing two salt solutions, or an alkaline and an acid solution, in order to bring about a reaction was the most widely used practice; more reactions took place in the wet than in the dry way. Heating a substance in air (for example, roasting lime to produce quicklime or heating mercury to produce its red calx, which we call mercuric oxide) corresponded to reaction in the dry way. Tables of affinities sometimes indicated that they referred to one or the other of these ways.

Information about chemical theory is easier to come by than information about chemical practice. Nonetheless, we should recognize that when chemists read tables of affinities, they had in mind not only the substances that would be produced but also the ways in which the appropriate reactions could be generated and controlled. So their explicit classification of substances through affinities was joined to an implicit classification of chemical operations. Chemical operations depend on chemical apparatus, some built for that purpose, and some available in any kitchen.

Chemists may have been sparing in describing their practice, but they were even more sparing in describing their apparatus. This may have been partly because most apparatus had changed little over the years, so readers could be expected to be familiar with it. Any eighteenth-century laboratory contained vessels for mixing substances in solution and vessels for mixing them in the dry way. The former group of instruments could include flasks, jars, and cooking pots. The latter group included crucibles and apparatus for bringing about sublimation, the transformation of a solid to a vapor and back again to a solid without the substance passing through a liquid phase. Reactions in the wet way could also involve distillation, and laboratories generally had a variety of apparatus for distilling substances. Distillation, calcination, sublimation, and other processes all depended on the application of heat, in varying intensities. Thermometers were not much used by chemists before the end of the eighteenth century, partly because they were not very accurate. Instead, experienced chemists observed the behavior of reactants to determine how intense a heat to apply, and for how long. They used water baths, in which substances or the vessels containing them were immersed in hot water, steam baths, sand baths, and a wide variety of furnaces. Much of the apparatus in use in 1700 had changed little in a century, and a surprising amount of it was not very different from the apparatus used by Arab alchemists in their heyday. Broadly speaking, the laboratory in the first half of the eighteenth century provided a stable but not a static environment in terms of apparatus and its uses. That situation was to change radically in the second half of the century, when chemistry benefited both from its own advances and from advances in the wider field of philosophical or scientific instrument making.

Heat and fire were chemistry's most powerful tools. We saw in Chapter 3 how for Stahl, heat was an instrument and fire a principle or substance. As the eighteenth century progressed, Rouelle's more complex view—in which heat could be here an instrument, there a principle, and sometimes both at once—began to transform chemistry and to bring the science of heat and the phlogiston theory to the forefront of chemical debate. Chemical theory underwent radical change, not for the first time and not without keeping one foot in its past, and chemical apparatus also were altered.

5 Different Kinds of Air

Words change their meanings as time passes. That is true in natural philosophy, in chemistry, and in all sciences, as well as in nonscientific language. *Air* is one word that has undergone radical change. Aristotle named Air as one of his four elements, compounded of the qualities hot and wet. Paracelsus had three elements, but air was not among them. Van Helmont had two elements, water and air, but for him air was not what we mean by a gas but was instead an originating principle. Van Helmont was, however, responsible for our word *gas,* which he coined from a Dutch or Flemish spelling of the Greek word *chaos.* The particles of gas, or of a gas, were in chaos, and gas could be a wild spirit because of its habit of escaping from chemical reactions. Van Helmont called "this Spirit, unknown hitherto, by the new name of Gas, which can neither be constrained by Vessels, nor reduced into a visible body."*

When air was produced in a chemical reaction, chemists tended to regard it as irrelevant but also as dangerous, since hermetically sealed apparatus could blow up from the pressure of contained air. Chemists got into the habit of making a hole in their apparatus so that air could escape without causing explosions. Robert Boyle measured what he called the "spring" of the air, but this was part of physical natural philosophy, not of chemistry. And, as Boyle observed, air was invisible, so that natural philosophers were inclined to "think [air] to be [the] next degree to nothing."† Air for John Mayow (1640–79) was essentially a receptacle for airborne particles of bodies, and through them manifested a variety of chemical properties. But although Mayow and a few other chemists did detect specific chemical properties in what we call gases (including our carbon dioxide), most chemists left them unaccounted for through all the years of alchemy and until the beginning of the eighteenth century. As chemists became aware that the atmosphere had a role to play in combustion, respiration, and in other reactions, they did not attribute this to the chemical properties of air but rather to substances that air could absorb and release ac-

*J. B. van Helmont, *Oriatrike* (London, 1663), 106.
†Robert Boyle, *Works,* ed. T. Birch (London, 1744), 5: 178.

cording to circumstances. Thus, as we saw, air for Stahl could soak up phlogiston as a sponge soaks up water and could also release it; air provided a physical environment in which some reactions took place.

In the early 1700s, the air was widely seen as just such an environment; and "air" and "the air" were one and the same thing. Chemists were not in the habit of regarding "airs" or "gases" as constituting a variety of chemical species. There was simply "air." One obvious reason for this was practical. Chemists could examine solids and liquids, exposing them to a variety of tests and seeing how they contributed to assorted reactions. Chemists had, however, no comparable way of examining air, and they came to view chemistry as the sum total of the reactions of solids and liquids, excluding gases. This view was easier to hold since the widely recognized failure of corpuscular and dynamical chemistry, the failure of physics to explain chemistry, and the resulting effort of chemists to achieve autonomy for their discipline. This effort led chemists to stress chemical qualities over physical properties like weight and to let physicists deal with air. Chemists generally did not examine air, and they did not try to weigh it. That does not mean that chemists did not weigh substances. They did a lot of weighing, and pharmacists and metallurgists did more. But weighing gases was outside their brief. In the *Encyclopedia* of Diderot and d'Alembert, readers were told that "the incoercibility of gases will remove them from our researches for a long time to come."*

By the time of the *Encyclopedia*, however, this had begun to change. One of the first and key sources of change was the invention by the Reverend Stephen Hales of a new instrument, the pneumatic trough. This instrument is important for what it made possible in the handling of air. The history of its invention and early use illustrates the difference there may be between the motives for inventing a device and the ways in which that device is used.

Hales was a botanist and chemist, as well as a Newtonian physiologist. He wrote a book with the title *Vegetable Staticks* (1727), investigating, as the title suggests, mechanical subjects like the pressure of sap in plants. But Hales went further, addressing chemical as well as physiological questions. He urged chemists to consider air chemically: "May we not . . . adopt this now fixt, now volatile Proteus among the chymical principles?"† He described an instrument for washing the air produced in the course of a chemical reaction. He wanted to get rid of impurities in the air by letting the air pass through water. Air passed from the reaction vessel, a retort with a long curved neck,

---

Encyclopédie (1757), 7: 520.

†Stephen Hales, *Vegetable Staticks* (reprint, London: MacDonald, 1969), 179–80. Proteus was a mythological sea-god who could turn himself into all kinds of fabulous shapes.

Stephen Hales's Pneumatic Trough

Hales was a country clergyman and, like many such clergy, took an interest in natural philosophy. His own bent was experimental, and he was successful enough to be elected to the Royal Society of London in 1717.

In what has become his best-known book, published a decade later, Hales was looking for ways to provide an essentially mechanical understanding of plant physiology, including the pressure of vegetable sap. Statics is the branch of mechanics that handles systems in equilibrium as opposed to dynamic systems, where bodies are in motion. The full title of his book is *Vegetable Staticks: Or, An Account of some Statical Experiments on the Sap in Vegetables: Being an Essay towards a Natural History of Vegetation. Also, a Specimen of An Attempt to Analyse the Air, By a great Variety of Chymio-Statical Experiments; Which were read at several Meetings before the Royal Society.*

Robert Boyle had given mathematical expression to the elasticity of the air, and Hales wanted to explore the nature of "that wonderful Fluid, which is of such importance to the life of Vegetables and Animals." He believed that air contained various kinds of particles, and in exploring the nature of air, he wanted to wash impurities out of it. That was the origin of an

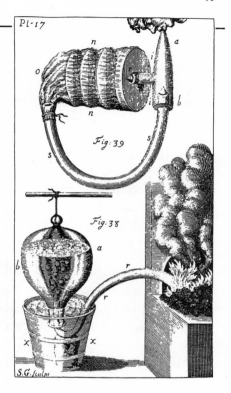

apparatus that, it soon appeared, could also be used for storing and isolating air, the so-called pneumatic trough. In this form, it became an essential tool for the chemical investigation first of air, then of different kinds of air, and, finally, of different gases and different chemical species.

■ Stephen Hales, *Vegetable Staticks* (London, 1727), plate 17, figure 38. Quote is from a 1969 reprint edition (London: MacDonald), xxiv.

through water in a trough and into an inverted vessel that had first been filled with water.

In devising this apparatus, Hales had coincidentally furnished an instrument for catching and holding air, which could then be subjected to various tests. Used in this way, the apparatus became known as the pneumatic trough.

Half a century after its invention, it became a staple of the chemical laboratory. It also became one of the key instruments in the reform of chemistry that we know as the "chemical revolution" (Chapter 6) because it was essential to incorporating a whole new state of matter, the gaseous state, into chemistry, alongside the already studied solid and liquid states. Once that step had been taken, it was possible to speculate and then to demonstrate that the gaseous state, like the solid and liquid states, could contain a variety of chemical substances. This was an enormous step, and it did not happen overnight.

Hales had shown that air could be contained, washed, and purified, and tested chemically as well as physically. This, however, did not lead him to think that there was more than one kind of air. Air for him remained "air," not one of a number of "airs." Other chemists would take that essential step, sometimes tentatively, sometimes, it seems, almost unawares.

Identifying, Measuring, and Multiplying Airs

The work of Joseph Black (1728–99) of Glasgow and Edinburgh came to be seen as a model for the chemical investigation of airs (in the plural) and as the first thorough characterization of a particular species of air. That was not, however, how his work was first received. In spite of a relatively short career in research, he became the most renowned professor of chemistry in the history of Edinburgh University. He began improbably with an M.D. thesis looking for a cure for bladder stones. Before the days of anesthetics and antiseptics, surgery was a dangerous and agonizing business, best avoided if at all possible. Samuel Pepys the diarist underwent an operation for "the stone," and every year on the anniversary of that operation gave thanks to God for its success. But perhaps there was a less drastic cure for the stone. Physicians and chemists knew that bladder stones could be dissolved in caustic alkali, but that was no help to physicians or their patients, since drinking caustic alkali would lead to a very painful death. One possible avenue of investigation was to seek a gentler way of dissolving bladder stones, one that would not harm patients. The search to make a medicine for the stone out of alkalis began with a tip from Van Helmont, and Black was working within a long-established tradition.

Black began by investigating the chemical properties of what he called magnesia alba, our basic magnesium carbonate, which is a much gentler alkali than caustic alkali is. He found that, on heating, magnesia alba yielded a gas and that the gas, when passed through a solution of limewater, turned that solution milky. This was not the case with atmospheric air. In the second part of his investigation, Black showed that when limestone was treated with mineral acids, it gave off the same air that the experiments on magnesia alba had yielded. He also showed that limestone on heating yielded quicklime and, once again, the same air as the previous experiments. Black concluded that he had demonstrated that there were two airs, or two kinds of air, atmospheric air and the air that had been held in combination in magnesia alba and produced from it. He showed that the latter kind of air was also produced in respiration, fermentation, and combustion. He called this air "fixed air" because it had been fixed in the solid magnesia and then liberated from it. His vocabulary here was not new, although his research was. Black knew that he was using a term that was already familiar in philosophy. Isaac Newton had speculated about the air contained in bodies. Stephen Hales, working in a Newtonian tradition, had described air as "now fixt, now volatile." But as we have seen, air for Hales was atmospheric air, the only kind of air there was. By using the old name of "fixed air" for it, Black made it possible to view his work in terms of earlier concepts.

Black not only characterized the new air, fixed air, qualitatively, in terms of its chemical behavior. He also added the dimension of quantitative analysis to the investigation, weighing the magnesia alba before heating and weighing the residue after heating. Then he made a step that to us is obvious, but one that had not been applied to gas chemistry before. He concluded that the weight of fixed air was equal to the loss in weight suffered by magnesia alba on heating. He was far from the first to perceive that a substance could be characterized by both the nature of its constituents and their proportions by weight, and that quantitative analysis could reinforce qualitative analysis. But he was the first to extend that perception to the study of gases. He repeated his experiment on magnesia alba and fixed air until he was satisfied that he had a reasonably consistent result—consistent to one part in 250. Modern examination of Black's balance, which survives in Edinburgh, shows that it was accurate to at least one part in 200. So Black had a pretty good sense of the accuracy and precision of his work. That is one reason later chemists looked back to Black's work with such respect. At the time, however, his work was mainly regarded as an important contribution to the chemistry of the alkalis. Black had shown that mild alkalis become more alkaline when they lose fixed air and that absorption of this gas restores the mildness of the alkalis. Alas, Black's work on magnesia alba did not produce a cure for bladder stones.

In characterizing fixed air, Black had measured it well enough, although indirectly, by attributing to it the weight lost by magnesia alba on heating. Henry Cavendish (1731–1810) showed how to measure gases directly and precisely. He was perhaps the shyest Fellow of the Royal Society of London and was undoubtedly the most brilliant and precise experimenter on the chemistry of airs. In another area of the sciences, his experimental determination of the gravitational constant has to rank as one of the greatest quantitative experiments ever. His experiments were distinguished by elegance, simplicity wherever possible, and the most careful attention to each detail of laboratory practice.

His first chemical publication was on three "airs" collected in an inverted jar over water. One of these airs was Black's fixed air. Cavendish repeated some of Black's experiments and added further quantitative observations of his own. He also investigated the inflammable air produced when zinc, iron, or tin is dissolved in hydrochloric or sulfuric acid. This air is our hydrogen. Cavendish found that the amount of inflammable air produced was the same when a given weight of one of these metals was dissolved in either of these acids. He reasoned that because different acids produced the same air, the air must have come from the metals and not from the acid. Metals, according to Stahl, were compounds of a calx with phlogiston, the principle of combustiblity. Cavendish speculated that the inflammable air produced might be phlogiston itself, and the action of acids released it from metals.

Cavendish investigated the chemical properties of this inflammable air, showed that it was distinct from other known airs, and managed to weigh the gas produced in a series of reactions. Weighing gases was not common in those days, and it was not easy, partly because gases were very light compared with the vessels in which they were contained, so that small percentage errors could lead to large errors in the result. Cavendish's quantitative results were remarkably good. When he investigated the composition of atmospheric air, he found not only dephlogisticated air (our oxygen) and an air that would not support combustion (our nitrogen) but also a tiny residue of an air that would not combine with anything he tried. This result was ignored for the next hundred years until the chemists who discovered the first inert gas, argon, rediscovered Cavendish's results and realized that his superb quantitative experimental results (although not his interpretation of them) had anticipated theirs. Cavendish did not, however, discover argon; a discovery involves awareness of the significance of results.

One area where Cavendish has as good a claim as anyone, although not a unique claim, is the discovery and determination of the composition of water. We shall return to Cavendish after considering the work of Joseph Priestley, the other great chemist of airs in eighteenth-century England. More than any other chemist, Priestley (1733–1804) was responsible for discovering new kinds of air. He was a doctor of theology, not of medicine and was the most prolific of British chemists in the eighteenth century, as well as the most public figure among them.

A dissenting minister (i.e., a clergyman who did not subscribe to the articles of faith of the established Church of England), Priestley denied the Holy Trinity and was widely though wrongly regarded as an atheist. Dissenters were not allowed to attend the universities of Oxford and Cambridge, and so they founded their own colleges and academies, with science or natural philosophy as an important part of the curriculum. Priestley became a teacher at one such academy, and he began to study the sciences in preparation for teaching them. He met Benjamin Franklin, English gentleman turned American sage, who helped him with his studies of electricity. Later, in Birmingham, he became a member of a remarkable informal group of natural philosophers and industrialists, whose members included James Watt, inventor of a new steam engine, Josiah Wedgwood, founder of the Wedgwood pottery, and Erasmus Darwin, physician and grandfather of Charles Darwin. He visited Paris and met the leading French chemists; he was subsequently elected as a foreign member of the Royal Academy of Sciences.

Priestley's politics were as unorthodox and radical as his religion. He was a democrat, and he supported the French Revolution. In England, horrified by revolution abroad and fearful of possible revolution at home, democracy was seen by the government as being every bit as subversive as supporting communism was in the United States during the McCarthy era of the 1950s. In 1791, on the anniversary of the fall of the Bastille, a mob encouraged by government-inspired propaganda broke into Priestley's house and destroyed its contents, including his laboratory. Priestley had the courage of his convictions. Years before the French Revolution broke out, he had written in the preface to his account of the chemistry of airs that "the English hierarchy (if there be anything unsound in its constitution) has . . . reason to tremble even at an air pump."* Critics of the French Revolution took him at his word and used chemical imagery to describe that revolution. Edmund Burke, the greatest parliamentary orator of his age and the most relentless critic of the French Revolution, had Priestley in his sights when he condemned the spirit of liberty: "The wild *gas,* the fixed air, is plainly broke loose: but we ought to suspend our judgment until the first effervescence is a little subsided, till the liquor is cleared, and until we see something deeper than the agitation of a troubled and frothy surface."† After the "king-and-country" antidemocratic mob attacked Priestley's home, Priestley left Birmingham and in 1794 quit England for America, where there was more sympathy for democracy.

*Joseph Priestley, *Experiments and Observations on Different Kinds of Air,* 2nd ed., 3 vols. (London, 1775–77), 1: xiv.
†Edmund Burke, *Reflections on the Revolution in France* (1790; reprint, Oxford: Oxford University Press, 1993), 8.

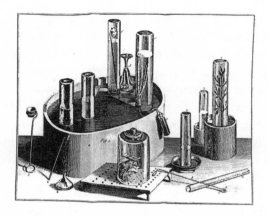

Joseph Priestley's Pneumatic Apparatus

Priestley, like Stephen Hales, was a clergyman and a natural philosopher. Unlike Hales, he was also very much a dissenter in matters of religion and politics, where his refusal to accept the doctrine of the Holy Trinity and his enthusiastic support for the French Revolution combined to make him loathed by the establishment. But he was essentially conservative in matters of chemical research, even though he made many new discoveries.

The pneumatic apparatus devised by Priestley is highly functional and also very simple. He used a variety of cylinders and jars, kitchen utensils, a gun barrel in which to heat substances, and only the simplest purpose-built apparatus. His friend Josiah Wedgwood made apparatus for him in stoneware and earthenware. The pneumatic trough really was a trough, with a shelf at one end. His eudiometers (which measured the goodness of gases) were simple jars, open at one end. In the illustration there is a wine glass (top center) and a medicine bottle (hanging on the trough's edge). Note the plants standing in water (at the right side of the trough) and in air over water (far right); Priestley contributed significantly to our understanding of vegetable respiration. See also the mice in the jar in the foreground. He used mice to determine the goodness of air for respiration until, to his considerable relief, he was able to develop strictly chemical tests and to avoid inflicting suffering on laboratory animals.

■ Joseph Priestley, *Experiments and Observations on Different Kinds of Air, and other Branches of Natural Philosophy, connected with the Subject,* 3 vols. (Birmingham, 1790), 1: frontispiece.

Priestley's chemistry was remarkably successful, and nowhere more so than in the realm of airs or gases. He showed great inventiveness in the laboratory and discovered more new gases than anyone of his generation. He has, however, had a bad press because he persisted in using the phlogiston theory to explain combustion long after most chemists had given it up. Both his contemporaries and subsequent historians have argued whether he should have been converted by the arguments put forward by opponents of the phlogiston theory. The phlogiston theory turned out to be wrong, but it was, as we have seen in Chapter 3, a good theory. It made sense of a great deal of experimental evidence and enabled chemists to make predictions and to propose new and fruitful experiments. Priestley was brilliant in using the phlogiston theory in this way. It is largely because of his successes that advocates of the theory of combustion that replaced the phlogiston theory toward the end of the century often referred to the new theory as the "antiphlogistic" theory.

In 1770 Priestley took up some of Hales's inquiries about air, prompted, he tells us, by living next to a brewery and noticing that fixed air was produced by the fermentation. One result of that early interest was his invention of soda water, water impregnated with fixed air. From there, he broadened his field of study. He used and modified apparatus invented by Hales and by Cavendish. He showed that the air in which a candle, wood, or alcohol has burned is also fixed air. In experiments, an enclosed volume of air in which a candle had burned until the flame went out would asphyxiate animals. Priestley's usual laboratory animal was a mouse, and he was careful to try to rescue and revive his experimental animals as soon as they became unconscious. But air that asphyxiated animals could be made healthful again by the action of growing vegetation. Priestley's interpretation of these results was that animal respiration, like combustion, released phlogiston into atmospheric air. When that air was fully saturated with phlogiston, no further combustion or respiration occurred, so flames and animals both expired. Growing vegetables, in contrast, removed phlogiston from the air, thus restoring the air's ability to support combustion and respiration. Plants and animals, the one removing and the other adding phlogiston to the air, existed in a mutually sustaining balance. Priestley thus far was elaborating on foundations laid by Stahl and by Black. But for him, this was just the beginning.

Priestley said that his discoveries owed more to chance than to design. In that case, he was a very lucky man. From fixed air, he moved to a series of other airs. Nitrous air, our nitric oxide, was one of his discoveries. He found that it produced acrid red fumes in atmospheric air, and if mixed with atmospheric air over water, there was a decrease in volume. He soon realized that such decreases in volume offered a way to determine how good an air was for breathing. He was delighted to find that he no longer had to asphyxiate laboratory mice to test the goodness of air.

Measuring the "goodness" of air for respiration was at first widely seen as primarily a medical question. Was the air in mines, in prisons, crowded rooms, or hospitals good for breathing? If not, cures might lie in ventilation or in chemical treatment. The goodness of air became, however, a chemical issue, largely as a result of Priestley's own work. Using the sun's rays brought to a focus by a burning glass, he heated different substances in a closed jar over mercury. One of these substances was the red precipitate, or calx, of mercury (mercuric oxide), prepared by the controlled heating of mercury in atmospheric air. But the calx, when heated, yielded an air that was markedly better than atmospheric air at supporting combustion and respiration. It could even revive recently asphyxiated mice. Priestley at first thought that this was his "nitrous air" (nitric oxide), but later he called it vital air, eminently respirable air, and, since it supported combustion, it was also evidently dephlogisticated air. Priestley was delighted with these experiments and announced "that there is no history of experiments more truly ingenuous [ingenious] than mine, and especially the Section on the discovery of dephlogisticated air."* Dephlogisticated air was one component of atmospheric air. Cavendish had shown that there was another component, which supported neither respiration nor combustion. Priestley called this component phlogisticated air.

    We should pause for a moment to take stock of gas chemistry by the mid-1770s, when Priestley completed his work on dephlogisticated air. Priestley empirically discovered more kinds of air, more different gases, than any of his contemporaries or predecessors did. His laboratory discoveries are many and important. But he could not be expected to interpret them as we do. He was working very much within the confines of qualitative phlogiston theory. The names he used for gases are derived partly from that theory and partly from observation of particular chemical properties. Those names alone do not discriminate between different kinds of inflammable air (e.g., hydrogen and methane), between different kinds of air that support combustion and are therefore dephlogisticated airs (e.g., oxygen and nitrous oxide, also known as laughing gas). Because Priestley did not always distinguish the airs in each group (inflammable, phlogisticated, etc.), there is a sense in which it is not reasonable to credit him with the discovery of oxygen, even though that was the dephlogisticated air that he obtained from red calx of mercury.

    Careful qualitative and quantitative experiments, such as those performed by Cavendish, did make additional discriminations between gases so that, for example, Cavendish clearly understood the difference between different inflammable airs. It is now time to return to Cavendish.

----

*Priestley, *Experiments and Observations* (1775), 2: ix–x.

Cavendish and the Production of Water

One of the mysteries in the history of chemistry is how seldom chemists blew themselves up while investigating novel substances and reactions. Hydrogen and oxygen (inflammable air and dephlogisticated air) can burn smoothly together, but they can also react explosively. Priestley used to carry small bottles of these two airs, and he entertained visitors by exploding the gases. Other chemists used static electrical discharges to ignite the mixed gases, with similarly dramatic results. Priestley noted that the combustion or explosion of inflammable air with atmospheric air involved the production of moisture. He also observed that the gas mixture diminished in volume. But it was Cavendish and not Priestley who decided to investigate the cause of that diminution.

Cavendish generated his inflammable air by the action of acid on zinc, mixed the inflammable air first with atmospheric air, weighed the mixture, ignited it, and weighed the remaining gases. There was always a weight loss and a destruction of the gas mixture's ability to support combustion. Cavendish confirmed what some other chemists had already proposed, that common air was a mixture of dephlogisticated and phlogisticated air, and that these airs were truly different substances, differing in more than their degree of phlogistication. He also observed the moisture or dew formed during the experiment and weighed it. After a series of experiments, he concluded that approximately two volumes of inflammable air combined with one volume of dephlogisticated air to produce their own weight of water.

His work was quantitative and based on the rule (held tacitly by chemists but thus far seldom used to promote discovery) that in chemical reactions, matter was neither created nor destroyed. Cavendish used the balance to guide his researches, to a degree not hitherto common. Joseph Black was in this respect his most eminent predecessor. We can interpret Cavendish's quantitative and qualitative results as showing that hydrogen and oxygen form water. Cavendish, working at first like Priestley within the phlogiston theory, interpreted his results otherwise. He decided that dephlogisticated air was simply water minus phlogiston, and inflammable air was either phlogiston, as Priestley had found, or water *plus* phlogiston. The combination of these two gases led to the canceling out of the plus and minus quantities of phlogiston and to the production of water. This meant that water, so long believed to be a chemical element, could still be one. Later on, Cavendish recognized that a new French theory of combustion and of gases made a different explanation possible.

Lavoisier: Phlogiston My Foe, or, the Oxygen Theory of Combustion

In the mid-nineteenth century, a dictionary of chemistry began with the assertion: "Chemistry is a French science, invented by Lavoisier."* The dictionary was of course also French and was making a nationalistic and polemical statement. The author was exaggerating to make a point. His point was that late eighteenth-century French chemists had brought about a major transformation of chemical science. And among these chemists, the most famous and the most innovative was the self-styled author of a revolution in chemistry, Antoine-Laurent Lavoisier (1743–94).

The story of Lavoisier's life and death is as dramatic as any in the history of science. Handsome, energetic, and wealthy, married to a beautiful young wife who assisted him in his researches, he soon became one of the most distinguished members of the Royal Academy of Sciences. But it was his misfortune that his rise to fame coincided with the destruction of the class of society to which he belonged. His most famous book, and one of the great landmarks in the history of science, was his treatise *Elements of Chemistry,* which was published in 1789. That was also the year in which the Bastille fortress and prison was demolished, the year of the outbreak of the French Revolution. For a while Lavoisier worked for and with that revolution, believing, as he wrote to Benjamin Franklin, that it could correct many abuses, if only it remained under the control of the right sort of people, people like him. The Revolution was soon beyond any such control, and it careered from popular uprising into the years of slaughter known simply as the Terror. Academies were suppressed, deemed elitist organizations, contaminated by association with the old regime. Lavoisier, whose wealth had grown through his activities as a tax farmer (one who collected taxes for the state), was imprisoned, condemned to death, and guillotined in 1794, along with many other tax farmers.

Lavoisier had been involved in a lot of scientific work for the state, including extensive work on gunpowder. He had also embarked on a reformation of chemistry, and by the mid-1780s he had convinced the other leading chemists in France that the phlogiston theory was no longer tenable and that a new theory, of which he was the principal architect and advocate, was preferable.

This replacement of an old by a new theory, framed by Lavoisier, is the essence of the chemical revolution. The revolution was complicated, and it will be useful to bear in mind some of the key ingredients in Lavoisier's work:

*Adolphe Wurtz, *Dictionnaire de Chimie* (Paris: Hachette, n.d.), 1: i.

1. His researches showed convincingly that the phlogiston theory did not stand up to quantitative studies, especially when air was included in these studies.
2. He provided an explanation of the heat and light produced in combustion as an alternative to the phlogiston theory.
3. He elucidated the role of oxygen in combustion and respiration and gave the substance that name. He fully incorporated the chemistry of gases into chemical science.
4. He developed a new theory of acids.
5. He provided a new definition of simple substances, based on laboratory practice.
6. He contributed to the formulation of a new nomenclature in chemistry, based on his understanding of simple substances and on a clear recognition that certain combinations of simple substances were stable, persevering through a series of reactions.
7. He developed a research method in which gravimetric analysis (which uses the balance to weigh and monitor reactants and products) was a way of monitoring reactions.
8. He devised new apparatus and improved old apparatus for the quantitative study of gases.

These contributions emerged over a period of some fifteen years of research, and they were not all separate in their development.

In 1772 Louis-Bernard Guyton de Morveau (1737–1816), a lawyer and chemist, showed that any metal that could be calcined would gain weight when it was transformed into a calx, and that the weight gain was fixed for each metal. When Lavoisier learned about these experiments, he thought that calcination might involve the fixation of air. A gain in weight was incompatible with the phlogiston theory, which interpreted combustion as the loss of phlogiston, not as the fixation of air. Later that same year, Lavoisier repeated experiments on the combustion of phosphorus and sulfur, confirmed that an acid was produced in each case, and once more noted an increase in weight. This too, he judged, involved the fixation of air. In 1774, Joseph Priestley came to Paris and met Lavoisier. He told him and other French scientists that he had obtained a new kind of air by heating the red precipitate of mercury (mercuric oxide). Lavoisier pursued experiments on the calx of mercury and found that, unlike Black's fixed air, the air fixed in the calx would support combustion. Priestley meanwhile had pursued his own researches and realized that the gas was dephlogisticated air. Lavoisier then put his findings and Priestley's findings together to show that the air was a constituent of the atmosphere, could combine with burning charcoal to form Black's fixed air, and was "eminently respirable air." What distinguishes Lavoisier's research from Priestley's is primarily his systematically used quantitative method, which we will consider in Chapter 6.

Lavoisier was working on several fronts at once. His experiments of 1772 showing that the combustion of sulfur and phosphorus led to the formation of acids were developed and reinforced by his work on calcination. By 1779 he had shown that the acids formed from sulfur, carbon, nitrogen, and phosphorus all contained eminently respirable air. He concluded that this air was essential to the formation of acids, and he called it *oxygen,* from the Greek for "acid generating." Oxygen was for Lavoisier the acidifying principle.

With the understanding that oxygen gas was a component of the atmosphere, he explored other combustion reactions. The inflammable air produced by adding zinc to a mineral acid would burn in oxygen. Lavoisier worked with colleagues to develop apparatus for burning gases and for storing and measuring them. In 1783, Cavendish's assistant and secretary visited the Academy of Sciences and told Lavoisier about Cavendish's synthesis of water from inflammable and dephlogisticated air. Lavoisier, using his new combustion apparatus, performed a continuous combustion of these airs and found that only water was produced. He concluded that water was a compound formed from oxygen and the inflammable air. Later, Lavoisier called the inflammable air *hydrogen,* meaning water producing or generating.

Lavoisier's interpretation of combustion as a burning substance combined with oxygen from the atmosphere made sense in terms of the weight changes observed when he included gases in his measurements. But the phlogiston theory did at least offer a qualitative explanation of the production of heat and light in combustion. In the old theory, phlogiston, as the principle of combustibility, was understood to be the cause of the phenomena of heat. Lavoisier, once he had decided to tackle the phlogiston theory, had to provide his own account for these phenomena. He found an explanation by postulating a matter of heat that had no weight, which could combine chemically with substances that did have weight.* He gave the name *caloric* to this matter of heat. If caloric could not be weighed, how could it be measured? Caloric was for Lavoisier a substance, and so it had to be conserved. Lavoisier wanted to

---

*The notion of imponderable fluid substances was common in the eighteenth century. Lavoisier's quantifying approach was distinctive. See Chapter 6.

demonstrate conservation of all matter, whether it possessed weight or not. As we shall see in the next chapter, his solution was to pursue experiments with a newly devised instrument, the ice calorimeter. New instruments, new methods, and a new language were the tools with which Lavoisier and his supporters achieved their chemical revolution.

6 Theory and Practice

The Tools of Revolution

Revolutions in general succeed either by force of arms or by peaceful persuasion, the force of argument and evidence. The French Revolution began with a mixture of physical force, the power of the people under arms, and argument, the rhetorical force exerted by advocates of the revolution. Force of arms should have no place in science, although coercion can impose new views on the practitioners of science, even in the face of clearly contrary evidence. In Russia under Stalin, the geneticist Lysenko (1898–1976) achieved enormous power and, with Stalin's support, was able to impose on Soviet scientists a theory of inheritance that had already been thoroughly discredited. Authority and patronage belong to the support structure of science, and they can exert less obviously coercive forces on scientists in any society, including our own, determining what research gets funded and published and what gets suppressed. They can determine and maintain what counts as good science. Natural philosophers in England who opposed Isaac Newton when he was president of the Royal Society of London were unlikely to gain academic or government appointments in science.

If Lavoisier and his supporters were going to achieve the revolution in chemistry that they began to regard as necessary, then unlike Stalin they would have to rely on force of argument and evidence. They needed to persuade chemists that the old phlogiston theory was wrong and that the new theory, including its oxygen-based accounts of combustion and acidity, was right. To do this, they needed to perform new experiments and reinterpret old ones. In carrying out their revolution, they devised new innovative apparatus and constructed a chemical language appropriate for the new ideas.

Language, the First Tool of Revolution

Language is a tool for communicating ideas and persuading others to accept them. It is also a tool for organizing knowledge and constructing rational arguments to reach conclusions. The chemical revolution was successful partly because its supporters managed to create a new language for chemistry and to make that language the normal language of chemical debate.

We have already seen that *dephlogisticated air* became *oxygen* and that part of the explanatory role of phlogiston was embodied in the new concept of the matter of heat, *caloric*. We have also seen that Lavoisier relied heavily on the principle of the conservation of matter, generally handled in terms of the conservation of weight, to show that reactants and products had all been identified and accurately measured. Another essential part of Lavoisier's practice was the recognition that chemical reactions were exchanges or regroupings of chemical components and that the nature of a substance was determined by its composition. All these aspects of the revolution required a radical reconstruction of the language of chemistry. And so the first tool of the chemical revolution was a new chemical nomenclature, a new language of chemistry.

Lavoisier himself clearly recognized the importance of language, of chemical nomenclature and of the rules governing that nomenclature. In the preface to his *Elements of Chemistry* (1789), he began by talking about the language of chemistry:

> When I began the following Work, my only object was to extend and explain more fully the Memoir which I read at the public meeting of the Academy of Sciences in the month of April 1787, on the necessity of reforming and completing the Nomenclature of Chemistry. . . . [But] while I thought myself employed only in forming a Nomenclature, and while I proposed to myself nothing more than to improve the chemical language, my work transformed itself by degrees, without my being able to prevent it, into a treatise upon the Elements of Chemistry.*

This statement is not quite accurate. Lavoisier knew very well what he was about. But why did he make the statement? What was the "Memoir" of 1787? What was the new nomenclature? And on what was it modeled? The new nomenclature did not simply spring into existence at Lavoisier's command. Although he took a major role in shaping it, the nomenclature was a product of the Enlightenment passion for thinking clearly, organizing knowledge, and classifying things. There were two main lines of thinking about the language of science that came together in the new chemical nomenclature. One can be traced directly to the work of the great Swedish naturalist Carolus Linnaeus; the other stems from French Enlightenment thinking about the language of science as it took shape in the work of Etienne Bonnot de Condillac.

---

*Lavoisier, *Elements of Chemistry*, trans. R. Kerr (Edinburgh, 1789 [1790]), xiii–xiv.

Linnaeus was the inventor of a new classification for plants. This botanical classification was embodied in a nomenclature of which several aspects were to prove important for the subsequent development of chemical nomenclature. First, his nomenclature was based upon the *observable characteristics* of plants (specifically, on the number of sexual parts of the flowers). Second, it was *binomial:* the name of each species had two parts, one identifying the species and the other identifying the genus to which the species belonged.* Third, it was in an *international* language, an important consideration for a discipline that transcended national borders, and doubly important for Linnaeus, working in Sweden; not many people outside Sweden read Swedish, then or now. Linnaeus chose Latin, traditionally the language of international scholarship. His system enabled him to arrange plants in groups according to their observable characteristics and to communicate unambiguously with naturalists in other countries. That represented a major advance over earlier plant classifications. Linnaeus developed his scheme from the 1730s through the 1750s.

Eighteenth-century reformers of chemistry, and several historians of chemistry since then, have represented the language of chemistry in Linnaeus's day as being full of ambiguities. Salts could be named according to their composition or according to the site where the mineral containing them was found; and of course there could be more than one such site, and so more than one such name. They could also be named after their discoverer, as in Glauber's salt, according to their taste, as in sugar of lead, and in many other ways. Even though eighteenth-century chemical nomenclature was not as chaotic as this suggests, there were enough problems to make the idea of reform seem reasonable. Linnaeus urged his compatriot Torbern Bergman, a professor of chemistry at Uppsala, to try to apply his methods to the classification of chemical substances, which could be viewed as species. Bergman in turn wrote about the project to the French chemist Guyton, and it was Guyton who brought the project to Lavoisier, and thus to the Royal Academy of Sciences in Paris.

The Academy was an institution with rules and bureaucracy. When faced with a problem, its members often resorted to the bureaucratic solution of forming a committee. A committee was appointed to look into the language of chemistry, with Guyton as one of its members and with Lavoisier as secretary. The secretary of a committee prepares the minutes, providing a written record of the committee's deliberations and recommendations. (Perhaps we should include committees as another of the tools of revolution.) In 1787,

*A genus is a group containing different species that all possess common structural characteristics distinct from those of any other group.

Lavoisier and his committee published their report, *Essay on Chemical Nomenclature*. Phlogiston was out, oxygen and caloric were in, and salts were named according to their composition, considered in two parts, metal and acid. Here was an analog to Linnaeus's binomial classification of plants. As a refinement, the ending of the acid part of a salt's name indicated how much oxygen it contained, as in calcium nitr*ite* and calcium nitr*ate,* where the -ate ending indicated a larger oxygen content than -ite. Some familiar names were retained, for example those of the common metals, but there were new names, corresponding to recently discovered or recently understood substances. Calling the breathable component of the atmosphere *oxygen* rather than *dephlogisticated air* meant that the new theory was embodied in the new language, and the old theory was excluded.

What of Condillac, the other influence on Lavoisier's idea of chemical nomenclature? Condillac had written that language was an instrument of reasoning, an instrument for rational analysis, and an aid to discovery. He had presented mathematics, the language of Newton's physics and astronomy, as the perfect language for science, and then had somewhat eccentrically identified that language with algebra. Mathematics in general, and algebra in particular, had clear rules governing the relations between its terms. The quantities in an algebraic equation could be moved around but not destroyed. Lavoisier's terms—copper and sulphate, mercury and oxygen—could similarly move from one combination to another, but they were not destroyed or diminished in the process. The idea of a chemical equation (although not in its modern form) is implicit in what Condillac says about algebra. Chemistry, if it could only find its right language, would become a science as rigorous in its deductions as mathematical physics.

There was one large question raised by the report on chemical nomenclature that Lavoisier and his colleagues submitted to the Academy. How could one identify the building blocks, the units to be manipulated in accounts of chemical reactions? Lavoisier's answer was *analysis,* a term familiar to chemists and to philosophers. At a stroke, he replaced all the old theories of the elements, and much although not all of the old chemistry of principles, with a chemistry of analysis leading to a provisional list of simple substances. It was a list based upon laboratory experience. That list, unlike philosophically based ones, was always open to change, since what cannot be decomposed and analyzed today may be analyzed tomorrow. Substances that had so far resisted analysis into simpler ones were the building blocks. Sulfur, iron, gold, and oxygen could not be decomposed or analyzed and so they, the simple substances of the laboratory, were building blocks for more complex substances. Copper sulphate, with its copper and sulfuric acid (itself a compound of oxygen and sulfur), and mercuric oxide, with its mercury and oxygen, were known experimentally to be compounds, and their names indicated their composition.

There were some problem areas. Chemists have always been guided by analogies between substances. That had led Lavoisier to argue that all acids were like those that he had successfully analyzed and thus had to contain oxygen, the acidifying substance or principle. The acids of carbon, sulfur, nitrogen, and phosphorus all fitted this picture. But what about the acid derived from common salt? By analogy, Lavoisier argued that it had to contain oxygen, and so he represented the substance that we call chlorine as a compound of oxygen and a radical that he named but could not isolate. Because chlorine contains no oxygen, we know that he was doomed to failure in his attempts to isolate its unknown radical; but he was not foolish to try. Overall, the new chemical language was what Lavoisier wanted, an instrument of discovery and a tool for persuading others of the truth of his new system of chemistry.

Balance and Gasometer

The instrument that Lavoisier used with the greatest success to demonstrate the truth of his system was the balance, an instrument with a beam pivoting on a central knife-edge, with a scale pan at each end. Chemical substances are neither created nor destroyed during reactions, and this truth can be shown to hold for any substance that has weight. We shall see later that Lavoisier needed a different instrument to try to show that caloric, the matter of heat, behaved as a simple substance in chemical reactions. This was a problem precisely because caloric had no weight.

The chemical species that Lavoisier recognized were simple or, if compound, were characterized by *constant composition*. The proportions by weight of one constituent to another were always constant in any pure compound. Thus, the proportion of oxygen to carbon in different samples of carbon dioxide was always the same. Constant composition could be demonstrated and confirmed by analysis, with the use of a balance. To use a balance in chemistry was nothing new. Many metallurgists, alchemists, chymists, chemists, mineralogists, and pharmacists had used the balance—some of them for the kind of precise diagnosis that Lavoisier wanted to find out the composition of substances; some of them used it to characterize substances. What was new was Lavoisier's use of the balance as a regulatory instrument. The essential point was to show that the weight of reactants was the same as the weight of products in any chemical reaction. In that case, he could be confident that he had not missed anything. If the weights before and after reaction were different, that would alert him to the need for further investigation. If there was a weight loss, then something had been lost in the reaction. An unexplained gain in weight showed that some other substance had entered into the reaction and needed to be identified. Lavoisier did not write modern chemical equations, but his idea of chemical reactions corresponded to the form of an equation. The two sides had to balance, and, in the chemical case, that balance involved simple substances and weights.

The balance was the most important instrument in Lavoisier's laboratory. He used two very different classes of balance, everyday ones that gave results good enough to show that he was on the right track and that enabled him to trace the overall nature of a reaction. But he also owned precision balances, made especially for him. He observed that, as "the usefulness and accuracy of chemistry depends entirely upon the determination of the weights of the ingredients and products both before and after experiments, too much precision cannot be employed in this part of the subject; and, for this purpose, we must be provided with good instruments."* Lavoisier put his money where his mouth was. He spent large sums on the construction of astonishingly precise balances. The best of them, according to one modern estimate, was capable of weighing to one part in 400,000. A comparable instrument had been made for Cavendish in England. Mechanical balances have never been more accurate than these masterpieces of the eighteenth-century's instrument makers' craft.

The balance might be astonishingly sensitive, but it was impossible to guarantee that the substances weighed had a corresponding degree of purity, and the different stages in the manipulation of reagents introduced their own minor errors. The result was that Lavoisier's precision balances were vastly more accurate than his experiments needed. Half a century later, Michael Faraday, one of the greatest experimental chemists ever, stipulated an accuracy of one part in 60,000, a high level of accuracy but much below that of Lavoisier's best instruments. Why then did Lavoisier and his instrument makers strive for and achieve such high precision? The answer lies in a variety of factors, powerful in combination. These include the pride of instrument makers in showing the best of what they could accomplish, the widely shared eighteenth-century passion for precision in science and measurement, and perhaps Lavoisier's own aims for chemistry as a science to rival the other physical sciences. Together, these factors turned Lavoisier's new chemistry into big science, with expensive and dramatic apparatus.

Lavoisier's gasometer was unquestionably the biggest, most expensive, and

*A.-L. Lavoisier, *Traité élémentaire de chimie*, 3rd ed., 2 vols. (Paris, 1801), 1: 333.

most impressive piece of chemical apparatus in his arsenal. It was the most expensive instrument in all of chemistry up to and beyond the end of the eighteenth century. The gasometer, which controlled and measured the uniform flow and volumes of the gases it dispensed, was an instrument made possible in theory and made necessary in practice by the emergence of gas chemistry as central to the chemical revolution. But, just as Lavoisier's precision balance was more precise than he really needed, so his gasometer was more elaborate and more expensive than laboratory practice required.

The chemistry of gases was central to Lavoisier's chemical revolution. One key experiment was the demonstration of the composition of water. We saw in Chapter 5 that Lavoisier carried out a continuous combustion of hydrogen and oxygen, confirming that water was the sole product of that combustion. To carry out the reaction continuously, he needed to measure and control the rate at which hydrogen and oxygen were supplied to the combustion apparatus. It was easier to do this by volume than by weight. Two volumes of hydrogen combine with one volume of oxygen to form water. Knowing the conditions of temperature and pressure under which the gases were held then made it possible to calculate the weights of the two gases that reacted to form water. In arriving at his results, Lavoisier began with apparatus similar to that used by Hales and Priestley before him. Then, during the 1780s, he went beyond them. Working with colleagues and instrument makers, he invented a true gasometer. The gasometers he used in his research were inexpensive and simple in design and, like his everyday balances, gave good approximate results. When it came to public demonstrations, however, he wanted apparatus that provided precise measurements. When one sees his precision gasometer, it is tempting to conclude that he also wanted apparatus that *looked* impressive and so promised a high degree of precision. Apparatus as well as language can have rhetorical effect.

Lavoisier's precision gasometer looked like the fruit of a marriage between Industrial Revolution engineering and physics, with nearly frictionless roller bearings, manometer gauges, dials and scales, as well as elaborately contrived chains. Each component had to be handcrafted by skilled workers. The resulting apparatus was enormously expensive, the Rolls Royce of chemical instruments, costing the equivalent of more than a quarter of a million dollars in today's money. *Two* such instruments were needed for the demonstration of the composition of water, one for hydrogen and the other for oxygen. Expensive or not, Lavoisier regarded his gasometer as precious and indispensable. Some experiments, he argued, simply could not be done without it. In Lavoisier's hands, chemistry had become big science—quantitative, precise, expen-

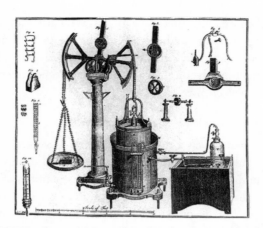

Lavoisier's Gasometer

The key instrument that Lavoisier used for public demonstrations of the composition of water was the gasometer. He claimed it was indispensable for all kinds of work in pneumatic chemistry. It was and is a very impressive piece of equipment. It stands nearly two meters high. Much of the gasometer is made of gleaming polished brass. The gauges are clearly the work of an instrument maker skilled in the production of precision apparatus. The Vaucanson's chain, which rides over the arm of the balance and supports the balance pan, is an exquisite piece of workmanship, and the arms sitting on the frictionless bearings atop the tallest column look remarkably like the heart of an ingenious pumping engine, a product of what was then a still-accelerating industrial revolution.

To use such an instrument or, as Lavoisier needed to do in demonstrating the composition of water, to use *two* such instruments was to exhibit control, authority, and virtual ownership of the field in which the instruments were used. No one else could afford to build gasometers like these. The results that Lavoisier in fact obtained using his gasometers were no better than what others soon obtained with much cheaper instruments; but, as the most expensive pieces of chemical apparatus anyone had ever seen, they showed that chemistry was big science, a science with prestige to rival physics and, perhaps, even astronomy.

■ Antoine-Laurent Lavoisier, *Traité élémentaire de chimie*, 2 vols. (Paris, 1789), plate 8.

---

sive, and demanding. Lavoisier observed: "It is an inevitable effect of the state of perfection to which chemistry is beginning to approach. This requires expensive and complicated apparatus: no doubt one should try to simplify them, but not at the expense of their convenience, and especially of their exactness."*

*Lavoisier, *Traité*, 1: 359–60.

For other chemists, Lavoisier's gasometers were, alas, simply too complicated and too expensive to build, so they sought cheaper and simpler alternatives. All converts to Lavoisier's chemistry did however accept, in the years of the consolidation of the chemical revolution, that gasometers were essential instruments of the new chemistry.

Ice and Fire: Ice Calorimeter and Blowpipe

Conservation of matter was a ruling principle for Lavoisier, and its demonstration was a regulative principle for him. He wanted to show, in every experiment, that nothing had been lost and that the quantities of different simple substances before reaction were the same as those after reaction, no matter what combinations or decompositions (separation into chemically distinct constituent parts) had occurred. For solids and liquids, weighing with a balance enabled him to demonstrate conservation. For gases, he had a choice between weighing directly and measuring volumetrically; from the volume and density of a gas he could then calculate its weight. There were, however, two sets of phenomena that he could not measure by weight or volume: the phenomena of heat and light.

We have seen that the phlogiston theory offered a qualitative explanation for heat in chemical reactions and that Lavoisier, in rejecting the phlogiston theory, had to provide an alternative explanation. Given his goal of making chemistry a demonstrative science, with a logic as rigorous as that of mathematics, he needed to be able to provide an explanation that could be demonstrated quantitatively as well as qualitatively. His solution was to say that there were two simple substances, light and caloric, which could both enter into chemical combination and be released from it. Neither light nor caloric, the old matter of heat and fire, had weight; that is, they were *imponderable* substances. The notion of imponderable substances was well established during the eighteenth century. Electricity was generally understood to consist of one or two imponderable fluids. Magnetism could be similarly explained. So could light and heat. Lavoisier extended the explanatory model by making two imponderable fluids, light and caloric, simple substances for the chemist. Light and caloric, however, were tricky. They could no more be captured and measured by volume than they could by weight. They were very much "wild spirits" that could not be coerced by the chemist in his laboratory. They possessed, in short, some of the same attributes that gases had possessed for chemists before the invention of pneumatic apparatus by Hales, Priestley, and others. That made it difficult for many chemists to accept them as chemical substances. If Lavoisier was to bring these supposed substances into his chemical system, and convince others that he was right to do so, he had to find a way to quantify them. He had to show that they were conserved through a series of chemical reactions, whether they entered into new combinations or escaped from old ones. If he could not do this, then his theory would fail to explain something that the phlogiston theory claimed that it could explain, and his revolution would be incomplete.

Lavoisier never did manage to measure light. But he was able to measure heat and to build it comprehensively into his theories of chemical change and physical state. Here, as elsewhere, he was making original contributions and also building on the work done by others. The study of heat had been a major preoccupation for eighteenth-century natural philosophers. Scottish researchers had led the way. Joseph Black, whose quantitative work on fixed air provided a fruitful model for later researchers, explored the phenomena of heat as well as those of ponderable substances. He observed that when ice melted, it absorbed heat without changing temperature. A mixture of water and ice would remain at the freezing point and absorb heat until all the ice had melted. Black concluded that since the heat had been absorbed without a rise in temperature, it must have combined with the particles of ice to form water. The heat was present, but it was not apparent as warmth, and so Black called it *latent* heat.* Following Black, natural philosophers had an explanation for the heat required to change a solid into a liquid, or a liquid into gas, without raising the temperature. Black measured the latent heat of the melting of ice, found that it was constant for a given weight of ice at the freezing point, and so by 1761 gave quantitative form to his theory of latent heat. In describing the latent heat as *combined* with the particles of ice, Black made it possible to claim the process as a chemical one. By measuring latent heat, he was pursuing the same quantitative approach that he had brought to his study of fixed air. Lavoisier had a ready-made home for his interpretation of the phenomena of heat as caused by combination with or release of the chemical matter of heat that he called caloric.

Heat had been a part of natural philosophy, the territory of physics; now it was being claimed as part of the territory of chemistry. That heat was related to change of physical state, from solid to liquid and from liquid to gas, was well known; substances melted or evaporated on heating. Lavoisier explained such transformations in terms of the acquisition of caloric. But what of the relation of substances which existed in chemical combination as part of a solid

*Black also found that it took different but specific amounts of heat to raise the temperature of the same weight of different substances by the same number of degrees; he thereby laid the foundations for the theory of specific heat.

substance and could be liberated from that combination as gases? Black's fixed air, Lavoisier's carbonic acid gas, was such a substance. So too was oxygen, part of solid metallic oxides but also capable of existing as a gas. Lavoisier interpreted the difference between fixed and free gases in terms of chemical combination with caloric. In mercuric oxide, for example, the substance oxygen was combined with mercury, whereas in gaseous oxygen, the substance oxygen was combined with the substance caloric. Gases were combinations of ponderable substances with imponderable caloric. Given this blending of physics and chemistry, it was fitting that when Lavoisier decided to measure the heat produced in chemical reactions, he did so by collaborating with the younger physicist Laplace.

The instrument that Lavoisier and Laplace invented to measure heats of reaction became known as the *ice calorimeter,* and its design rested squarely on Black's theory of latent heat. It seemed crude compared with Lavoisier's great gasometer, but its design was conceptually elegant. It consisted of a large bucket with a lid which was placed inside two other nesting buckets, both of which could be filled from the top and had faucets at the bottom. At the beginning of the experiment, the two outer nesting buckets were filled with ice, and any water in the inner of these two vessels was run off through the faucet, which was then closed. The ice in the outer vessel would melt gradually, but as long as some ice remained, its temperature would not change from the freezing point. That meant that the inner, or middle, vessel received no heat from the outer layer, since both began at the freezing point; in other words, the middle vessel was thermally insulated from the outside environment in the laboratory. Next, the chemical reaction under investigation was produced in the inside bucket. It might involve mixing two substances, and a modification to the apparatus made it possible to carry out gaseous as well as solid and liquid reactions. The heat released during the reaction would melt ice in the middle vessel, producing water that could be run off and measured at the end of the experiment. Again the principle of latent heat came into play, enabling Lavoisier and Laplace to calculate from the weight of water melted just how much heat had been released during the reaction. Not merely reactions between nonliving substances but also reactions in living animals could be investigated. Respiration, for example, in a guinea pig, was seen as a kind of combustion in the lungs, and it produced heat that would melt ice, and so could be measured.

Alas, thermal phenomena in chemical reactions are more complicated than allowed for in the simple model of combination with caloric or release of caloric. Not until the development of chemical thermodynamics and thermo-

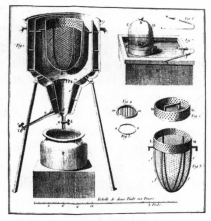

The Ice Calorimeter of Lavoisier and Laplace

When Lavoisier invited the younger Laplace to collaborate with him in a quantitative study of heat, the invitation was a great compliment to the younger scientist. The results of their joint research were presented to the Royal Academy of Sciences in 1783. The collaboration was symbolic of the union of chemistry with physics and a landmark in the quantification of what Lavoisier regarded as an imponderable substance, that is, a substance without weight. Both scientists were committed to the search for precise measurements, at a time when instrument makers were producing apparatus of unprecedented accuracy.

The measurements that Lavoisier and Laplace obtained with their instrument, which came to be known as an *ice calorimeter*, were in fact not very accurate; indeed, they were crude when compared with the precision that Lavoisier could obtain with his best balances. They were also crude in comparison with the astronomical data with which Laplace constructed his study *Celestial Mechanics* (1799–1825). But they did amount to a statement about the quantification of the study of heat as an alternative to the phlogiston theory that Lavoisier discredited, and they helped to persuade Laplace to give his significant support to Lavoisier's new system of chemistry.

■ Antoine-Laurent Lavoisier and Pierre Simon Laplace, "Mémoire sur la chaleur," *Mémoires de l'Académie Royale des Sciences* (1780, published 1784), plates 1 and 2.

chemistry in the late nineteenth century (see Chapter 12) did theory truly make sense of the observations. But Lavoisier, aided by Laplace, was striving to expand the reach of quantification in chemistry. Not for the first time, however, theory proved inadequate for interpreting experimental data. But the research program was a clear, rational, and disciplined one, resting on the best of modern science.

Lavoisier's quantitative method was crucial to the creation and to the success of the chemical revolution. Chemistry, however, is a science of qualities as well as of quantities. Chemical analysis has to be qualitative as well as quantitative. Chemists need to know what substances they are dealing with, as well as how much of each of those substances is present. Lavoisier recognized the importance of traditional operations for separating substances that were mixed rather than combined, including solvent extraction, crystallization, and fractional distillation (where substances with different boiling points are distilled at those different temperatures from a mixture).

Lavoisier's operational definition of a simple substance as one that had not yet been decomposed was an invitation to chemists to decompose whatever they could and to identify or discover as many undecomposed substances as possible. Heat was a powerful agent both for separating substances from one another and for decomposing a substance, either on its own or in combination with another substance. Furnaces of all kinds were traditional sources of heat for such purposes, as was the blowpipe, an instrument that Lavoisier took for granted, and so discussed only briefly in his writings. The blowpipe turned out to be one of the most powerful instruments for the identification of new simple substances.

The blowpipe is a narrow tube through which a stream of air can be blown. When applied to a flame, it produces a fine jet at a high temperature. Jewelers, glass workers, and craftsmen working with metal had used the instrument since antiquity, and it had remained a tool of skilled artisans for thousands of years before it was used in chemistry. In the eighteenth century it came into wide use as a specifically chemical instrument, first in Sweden and then, over the next half-century and more, throughout the rest of Europe.

Once it entered chemistry, the blowpipe proved to be a most delicate instrument for the qualitative analysis of mineral ores, revealing the presence of minute quantities of metal in very small samples. Chemists typically worked with samples the size of a mustard seed, and in those samples could detect even half a percent of a particular metal. This was far more sensitive than analysis in solution, the "wet way." The blowpipe led to the discovery of several metals during Lavoisier's lifetime, including nickel, manganese, molybdenum, and tungsten. Because these metals could not be decomposed, Lavoisier duly listed them in his table of simple substances.

Big Science, and New Borders for Chemistry

Robert Boyle had made chemistry part of the new and eminently respectable natural philosophy based on corpuscular philosophy. His explanations for chemical phenomena were ultimately compatible with corpuscular science. This made chemistry theoretically a part of what we would call physics. Chemistry as a practical science was something else, and Boyle contributed significantly to chemistry's independence from natural philosophy, at least in the realm of laboratory practice. His classification of chemical substances, based upon experiment, also distinguished chemistry from the rest of natural philosophy. And, as we have seen, Boyle saw no incompatibility between alchemy and his chemistry or chymistry, or between alchemy and atomism. The same is broadly true of Isaac Newton.

The failure of the corpuscular program in chemistry, and the similar failure of Newtonian chemistry, was reinforced by developments in Germany and France. The emergence in Germany of a chemical community whose members claimed independence from medicine and from physical natural philosophy contributed to the autonomy of chemical science. The stress upon experience by the chemists of the French Enlightenment had a similar effect. Chemistry became a science in its own right. Many historians and chemists have claimed that it did so even before the emergence of a recognizable science of physics.

Where do Lavoisier's chemistry and his chemical revolution fit into these developments? His definition of simple substances, based upon laboratory experience, argues for an independent science of chemistry. So does his use of qualitative analogies in classifying substances. Similarly, viewing chemical composition as the key to a classification of minerals and other compounds makes a case for the independence of chemistry. But Lavoisier was also acting and arguing in a way that aimed to give chemistry parity with physical science. He stressed that chemistry should strive for rigor in its arguments, just like mathematics and physics. He did more than any other chemist of the eighteenth century to transform chemistry into a quantitative science, in which sophisticated instruments were used to measure the quantities. His development of expensive and imposing apparatus bolstered the notion that chemistry was comparable with Newtonian physics and astronomy. Like them, it was big science. His collaboration with Laplace in their work on caloric showed that chemistry and physics were capable of being mutually supportive. The important role that he gave to caloric took what some still regarded as a part of physics and made it part of chemistry, as Black's work on latent heat had suggested. In short, Lavoisier helped to transform the theory, the status, and the boundaries of chemical science. The mid-nineteenth century assertion that chemistry was a French science, invented by Lavoisier, did have a grain of truth in its overstatement.

7 Atoms and Elements

Lavoisier's system of chemistry depended on the consistent use of a new experimental method, at least, one that had not previously been systematically applied in chemistry. Lavoisier asserted that he would be guided only by experimental evidence, by the facts of experience. Besides the evidence of the senses, the most convincing chemical evidence for Lavoisier was what could be measured and built into a quantitative system. Quantification opened the way to mathematical deductive reasoning. Lavoisier believed that mathematical reasoning based on measurement could make chemistry as rigorous and prestigious as Newtonian physics. Measurement of any substance that had weight meant the use of the precision balance or the conversion of measurements of volumes into the corresponding weights. We have also seen his work with Laplace on the measurement of heat, using the ice calorimeter to try to bring quantitative rigor to the study of caloric. Conservation of matter was the key.

For Lavoisier, weights were not simply numbers to tally. They were ways of regulating, shaping, and validating chemical experiments and theorizing. Chemical analysis and weighing went hand in hand. Together, they gave Lavoisier a way to replace the old ideas about elements with a new concept of simple substances of the laboratory. Chemical theories based on anything other than experimental evidence were, for Lavoisier, simply worthless, and that went for every theory of the elements that had been proposed by his predecessors.

Talking about true elements as indivisible atoms was metaphysics, not chemistry. Lavoisier insisted on making that point:

> If, by the term *elements,* we mean to express those simple and indivisible atoms of which matter is composed, it is extremely probable that we know nothing about them; but, if we apply the term *elements,* or *principles of bodies,* to express our idea of the last point which analysis is capable of reaching, we must admit, as elements, all the substances into which we are capable, by any means, to reduce bodies by decomposition.*

*A.-L. Lavoisier, *Traité élémentaire de chimie,* 3rd ed., 2 vols. (Paris, 1801), 1: xvii.

As we saw in Chapter 5, that did not mean that substances not yet decomposed were truly elementary. They might very well be compounds, and chemists might, one day in the future, be able to decompose them. Lavoisier stressed the provisional nature of his list of "elements": "Thus, as chemistry advances towards perfection, by dividing and subdividing, it is impossible to say where it is to end; and these things we at present suppose simple may soon be found quite otherwise. All that we dare venture to affirm of any substance is, that it must be considered as simple in the present state of our knowledge, and as far as chemical analysis has hitherto been able to show."* Lavoisier had a well-founded hunch that some substances he had been unable to decompose would prove to be compound, although he could not yet say what their constituent elements were. The alkalis, soda and potash, were substances that he was sure were compounds, but he had to list them as simple because he could not yet decompose them. In their case, Lavoisier's hunch was right. In another instance, that of chlorine, his certainty that it was compound was to prove unfounded. Later chemists showed ingenuity and spent a good deal of effort trying to decompose Lavoisier's "elements." Sometimes, as we will see, they succeeded, discovering new elements. Sometimes they failed.

Note that Lavoisier, like Boyle before him, distinguished between ultimate atoms and the undecomposed substances of the laboratory, what we call chemical atoms and physical atoms. There was no accessible way in Lavoisier's system to connect ultimate atoms with chemical elements. Sticking to the facts in Lavoisier's way meant avoiding making any connection between atoms and elements. There was, philosophically, a profound distinction between them. For the next two generations, chemists were to argue about that distinction. Some were sure it was necessary, some were equally sure that it was unnecessary and simply wrong.

John Dalton, Atoms and Elements: The Birth of a Chemical Atomic Theory

John Dalton (1766–1844) was the most influential of those who argued that chemical atoms were also physical atoms. He was the inventor of a truly chemical atomic theory, where atoms were the least part of chemical elements. He also discovered a series of numerical laws and rules expressing the rules governing chemical combination.

Dalton's father was a weaver and a Quaker, a dissenter rather than a member of the established church in England. As a Quaker, Dalton, like Joseph

*Ibid., 1: 194.

Priestley before him, was unable to attend either of the ancient universities. He was largely self-taught in science, and in 1793 began to teach mathematics and natural science in a dissenting academy in Manchester. He went on to become a major research chemist. By the time he died in 1844, he was internationally renowned and one of Manchester's most famous citizens.

Dalton's interest in science began not with chemistry but with meteorology, the science and study of weather. He was fascinated by the changeable and rainy climate of England's Lake District, where he was born and where he taught before moving to somewhat less rainy Manchester. From making meteorological observations, he began to think about and experiment on vapor pressure, the pressure exerted by gases or vapors. He also gave some attention to the composition of the atmosphere, mainly to nitrogen and oxygen. And he began to look at the solubilities of different gases in water. All these interests led him in 1801 to what we now know as *Dalton's law of partial pressures,* which states that in a mixture of gases, each gas exerts its partial pressure, the same pressure that it would exert if it alone were present. The total pressure exerted by a mixture of gases, like the atmosphere, is the sum of the partial pressures of the different gases present. It is as if the particles of one gas ignore the particles of other gases present in a mixture. Why would they behave like that?

Here, Dalton found inspiration in his reading of Isaac Newton. He found what was for him a crucial passage in Newton's *Opticks,* and he copied it into his own notebooks. "It seems probable to me," wrote Newton, "that God in the Beginning form'd matter in solid, massy, hard, impenetrable, moveable Particles . . . . While the Particles continue entire, they may compose Bodies of one and the same Nature and Texture in all Ages."* If the particles changed, if they broke in pieces or were worn down, then the substances that they composed, for example water and earth, would change their nature. What Newton meant by these statements is one thing and has been widely debated. What Dalton came to understand by them was another. He concluded that different simple substances or elements, such as oxygen or nitrogen, consisted of atoms. All atoms of a given chemical substance were identical with all the other atoms of that substance and different from the atoms of any other substance. That meant that Lavoisier's elements had to be considered as consisting of ultimate and indivisible atoms.

In the gaseous state, according to Dalton, atoms of the same kind repelled one another. As he put it, each atom "supports its dignity by keeping all the rest . . . at a respectful distance." That was why gases mingled evenly in the at-

*Isaac Newton, *Opticks* (New York: Dover, 1952), 400.

mosphere, instead of those with heavy atoms sitting at the bottom, near the ground, and the lighter ones occupying a higher place. Chemical analysis and synthesis were merely the separation or reunion of atoms. "No new creation or destruction of matter is within the reach of chemical agency. We might as well attempt to introduce a new planet into the solar system, or to annihilate one already in existence, as to create or destroy a particle of hydrogen."*

At this point, it is easy to imagine what Lavoisier, if he had been able to escape the guillotine, might have had to say to Dalton. Lavoisier's views about atoms and elements were very different from Dalton's. Lavoisier insisted that ultimate atoms were metaphysical and that we have no knowledge of them from experiment. He asserted that ultimate elements were also metaphysical. What he was willing to call chemical elements were the last products of analysis, and as our powers of analysis grew, he argued, the number of elements would change. Chemical elements had nothing to do with atoms, and chemical atoms were nothing but a fiction.

Lavoisier, given his philosophy and the chemical theories that he faced, was right to take the line that he did. But Dalton made a crucial breakthrough. He believed that the ultimate atoms of different chemical elements were distinguished by having different *weights.* All the atoms of a given element had the same weight. These were assumptions. Dalton made one further assumption. Lavoisier and others had shown that different chemical compounds could be characterized by the different combining weights of their constituents. Each compound had a constant composition.† There was, for example, a constant ratio between the weight of mercury and the weight of oxygen which produced mercuric oxide. Similarly, the ratio by weight between sulfur and oxygen in copper sulphate was constant, and it was different from the constant ratio between the same elements in copper sulphite. Dalton suggested that these ratios were the ratios of the relative weights of the atoms. Thus, if we agreed to take the relative weight of one substance as a reference, we could call that the atomic weight of the reference element. Then, by means of quantitative analysis, we could arrive experimentally at the atomic weights of every other element. But we could not do this without making more assumptions.

Dalton proposed, first, that like atoms repelled one another. That meant

*John Dalton, *A New System of Chemical Philosophy* (London: Peter Owen, 1965), 162–63.

†This depends on what counts as chemical combination. If metallic alloys and liquid solutions are combinations, as some chemists around 1800 believed, then obviously not all compounds are characterized by constant composition. Claude-Louis Berthollet was among the leading chemists who adopted the view that solutions and alloys were chemical compounds, and he rejected Dalton's laws of constant combining proportions (for these laws, see below). Dalton took the view that if composition was not constant, then the substance was not a chemical compound.

that elements, which were made up of like atoms, could not exist in diatomic molecules, where a molecule is a chemical combination of atoms. So oxygen and nitrogen particles in the atmosphere had to be single atoms of oxygen and nitrogen. He was wrong; we know that oxygen and nitrogen are diatomic gases, that is, that their molecules have two atoms apiece. But he needed working assumptions to bring order out of chaos. Both his studies of partial pressures and his search for simplicity in nature favored the rule that like atoms repel one another. And simplicity was something that Newton had recommended. Nature, according to Newton, is very simple.

Dalton's second assumption or rule was that when only one combination between two elements can be obtained, we should assume that its molecules consist of one atom of each element, unless, he allowed, "some cause appear to the contrary."* Ammonia would then be made up of one atom of nitrogen and one of hydrogen. But this is wrong, since we know that ammonia contains three atoms of hydrogen and one of nitrogen. If we take hydrogen as our reference standard and give it the atomic weight of one, then Dalton's formula means that the atomic weight of nitrogen is roughly 4.7. It isn't; we know that it is 14, three times as much as Dalton's assumption indicates. Dalton was wrong, but his argument was reasonable and his experiments were sound. They were also carried out, as Priestley's had been, with very simple and readily available apparatus. In spite of Lavoisier's claim that expensive apparatus was now essential, it was clearly possible to obtain important results without spending a fortune on instruments and apparatus.

When there were two combinations between the same elements, Dalton proposed that the simpler one should be assumed to be a 1:1 combination and the more complex a 1:2 combination. Thus, the gaseous combinations of carbon and oxygen would be carbon monoxide (one atom of carbon plus one of oxygen) and carbon dioxide (one atom of carbon plus two of oxygen), which is right.

Dalton's combining rules can be generalized in what is now known as *Dalton's law of multiple proportions*. When two elements combine in a series of compounds, the ratios of the weights of one element that combine with a constant or fixed weight of the second are small whole numbers. That follows from Dalton's account of the combining ratios of atoms. Because the molecular weight of a compound is proportional to the number of atoms of each element that forms part of that compound, and because atoms combine in the ratio of small whole numbers, combining weights will similarly be in the ratio

*Dalton, *A New System*, 167.

of small whole numbers. We can easily see how the law of multiple proportions applies to a series such as that of the oxides of nitrogen, N_2O, NO, and N_2O_4, or of the combinations of hydrogen and oxygen, H_2O and H_2O_2, using modern formulas which enable us to predict the combining weights in each case. The oxides of iron, FeO, Fe_2O_3, and Fe_3O_4, make up another series that obeys the law of multiple proportions.*

Dalton proposed a series of chemical symbols corresponding to each element or, in different contexts, to one atom of each element, and drew groups of these symbols to represent molecules. This meant that chemical formulas could represent the number of atoms in a compound, not merely the relative combining weights.

Let us take stock for a moment. Dalton was dealing with *combining weights* and with a *chemical atomic theory* that he invented. That led him to propose and determine a system of *atomic weights*. He both assumed and strengthened the *law of constant composition,* that each chemical compound has a fixed composition. He showed that chemical elements (whether provisionally defined, following Lavoisier, or taken as ultimate) combine in predictable proportions by weight, and he suggested rules for interpreting different compounds formed by the same two (or more) elements. This corresponds to the *law of multiple proportions.*

Most chemists could agree about combining weights, constant composition, and multiple proportions. Most chemists were also delighted at the possibility of a system of formulas that would indicate the relative amounts of elements in compounds. Such formulas were a wonderful help with classification, as well as provided a valuable shorthand that made clear the difference between nitrites and nitrates, sulphites and sulphates. Formulas, combining weights, constant composition, and multiple proportions offered the possibility of a research program that would determine combining weights and formulas so as to bring order to chemistry. Dalton's discoveries and rules helped nineteenth-century chemists to find order and predictability in the richness of chemical experiments, in the same way that the search for tables of affinities had served many eighteenth-century chemists. Dalton's contributions were among the most important in the history of chemistry.

---

*Dalton's simplicity rules here give him incorrect formulas. The simplest combination of hydrogen and oxygen, water, should be HO according to Dalton, whereas the correct formula is H_2O; hydrogen peroxide would have been HO_2 for Dalton, but it is in fact H_2O_2. Similarly, the simplest compound of nitrogen and oxygen was NO for Dalton, but it is N_2O for us. Although Dalton's simplicity rules sometimes led him to wrong formulas, they were entirely compatible with his laws of combining proportions.

John Dalton's Atomic Symbols

Dalton's table of chemical elements is the first in which chemical atoms and chemical elements appear in a one-to-one correspondence. Chemical elements for Dalton, as for Lavoisier, were substances that could not be decomposed into simpler substances. But Dalton went beyond Lavoisier, claiming that each chemical element was made up of identical and indivisible atoms, which were the element's smallest parts. Chemical elements were distinguished from one another by their chemical qualities and also by the different weights of their constituent atoms. Dalton did not claim to know the absolute weights of atoms, but made claims about their *relative* weights.

In Dalton's table, the first element is hydrogen, and the relative weight of an atom of hydrogen is 1. The fourth element is oxygen, with relative weight 7. That means that an atom of oxygen is 7 times heavier than an atom of hydrogen. Dalton obtained that result by observing that an atom or the least part of steam or water (no. 21 on the table) contains seven times more oxygen by weight than it does hydrogen; we call the least part of steam or water a molecule. Dalton further assumed that the molecule of water, since it was the simplest compound of oxygen and hydrogen, contained one atom of each element. If he had had more accurate data and

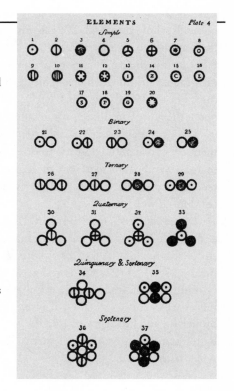

made the same assumption, he would have concluded that an atom of oxygen was 8 times heavier than one of hydrogen. Using today's knowledge that a molecule of water contains two atoms of hydrogen and one of water, and that each molecule has eight times more oxygen than hydrogen by weight, we arrive at our figure of 16 for the atomic weight of oxygen.

■ John Dalton, *A New System of Chemical Philosophy* (London, 1808), plate 4.

What Lavoisier had said about the unknowability of ultimate elements and the distinction between chemical species and physical atoms continued to trouble a lot of chemists long after the publication of Dalton's ideas in the early 1800s. Was it possible to reject Dalton's chemical atomic theory while adopting everything else that he offered? A significant minority of Dalton's contemporaries answered with an emphatic yes. Take the combination between hydrogen and oxygen. Everyone could agree that eight parts by weight of oxygen combined with one part by weight of hydrogen. Why not simply say that eight parts by weight of oxygen were *equivalent* to one part by weight of hydrogen, and then present that empirically determined result by saying that if one took the *equivalent weight* of hydrogen as 1, then the equivalent weight of oxygen was 8. Formulas could then represent the number of equivalent weights involved in a compound; carbon dioxide would have one equivalent of carbon to two equivalents of oxygen. For practical purposes, at least in the early nineteenth century, it made no difference whether chemists used a system of atoms or of equivalents. Later on, as we shall see, it did make a difference. But when Dalton was awarded one of the first two Royal Medals from the Royal Society of London, the president of the Society, Humphry Davy, made it clear that the award was for Dalton's laws of combining proportions rather than for his hypotheses about atoms. Davy went on to describe Dalton as the Kepler of chemistry. Johannes Kepler had come before Newton and had made great contributions to astronomy, but in England Newton was regarded as the one who finally made sense of the laws governing the planets. Dalton, by implication, would take second place to the Newton of chemistry.

Humphry Davy and the Voltaic Pile: Laws and Order

Humphry Davy (1778–1829) was a self-made man, a woodcarver's son who later, as the leading British chemist of his day, became president of the Royal Society of London. He made chemistry fashionable in London. Carriages queued up to bring people to his lectures or to read bulletins of his health when he was ill. He was constantly probing the nature and number of chemical elements, and in doing so he made the most dramatic use of an instrument invented by Alessandro Volta. Davy kept returning to questions about the existence and nature of atoms. And he, even more than John Dalton, saw himself as the Newton of chemistry, the one who brought laws and order into the chemical laboratory.

Before 1800, electricity meant static electricity, generated by friction. It could be stored in jar-like condensers, and a number of these condensers could be discharged simultaneously, like an artillery battery, producing a very hefty shock—up to half a million volts. The sparks from such discharges could ignite gas mixtures and decompose relatively small samples of some substances. Then in 1800, Volta published a description of a new piece of apparatus, the *electric pile.* It was called a pile because it consisted literally of a pile of alternating disks of metals and blotting paper moistened with a salt solution. It was also soon called a battery, by analogy with the battery of condensers used to store static electricity. When the top and bottom of the pile were connected, a continuous current flowed. It was not immediately obvious that this continuous kind of electricity was the same as frictional electricity. But it was clear that chemical action and the electrical action of the pile were connected in some fundamental way. Here was an immediate challenge, one that Davy described as "an alarm bell to the slumbering energies of experimenters in every part of Europe."*

The original form of the pile, a column of disks, soon gave way to a trough containing a salt solution, into which metallic plates were dipped, maintaining the original alternation of metals and salt solution but in a different form. Researchers observed that when the current passed through salt solutions, gases evolved and metals deposited on the wires or plates dipping into the solutions and connected to the opposite ends of the pile. As the source of a continuous electric current, the new form of the instrument was a recognizable ancestor of today's electric batteries. Since it produced chemical decomposition, it was an instrument of chemical analysis, an addition to chemists' arsenal of tools for breaking down compounds. It was also, apparently, an instrument that could be made more and more powerful, given that there was no theoretical limit to the size or number of metallic disks. A bigger and more powerful battery would be a more powerful instrument for chemical analysis.

Davy was the most successful of those who accepted the challenge implicit in Lavoisier's definition of elements as the last products of analysis. If he could decompose one or more of Lavoisier's elements, then he would have discovered new ones. As Lavoisier had observed, there was no telling where this process of discovery through decomposition or analysis would lead. Davy aimed to find out. He produced a series of ever more powerful electric piles. In 1806 he lectured to the Fellows of the Royal Society on the chemical agencies of electricity. He concluded, from a beautifully controlled and reasoned chain of electrochemical researches, that chemical affinity and electrical attraction were different manifestations of the same power of matter. He suggested that substances differed chemically in their response to an electric current, in a way that made it possible to rank them in a series according to their electropositivity or electronegativity. Metals, for example, were electropositive, while oxygen was electronegative. Electropositive substances were those that, in the course of electrolysis, went to the negatively charged pole, while elec-

*Royal Institution of Great Britain, H. Davy MS 1.

tronegative substances went to the positively charged pole, since opposite charges attract one another.

If chemical affinity and electrical attraction were ultimately the same, then chemical attraction was a form of electrical attraction, so that electropositive and electronegative substances would attract and combine with one another. Their combination could be overcome by a more powerful electrical attraction (e.g., from a strong electric battery) and electricity would therefore be able to produce chemical decomposition. Davy applied this thinking to the fixed alkalis, soda and potash. He knew, following the work of French chemists, that ammonia was an alkali and that it was a compound of hydrogen and nitrogen. By analogy, might not the caustic alkalis soda and potash also be compounds? He knew that Lavoisier had suggested that these alkalis were compounds, but Lavoisier, unable to decompose them, had listed them as elements, at least for the time being. Davy used a powerful voltaic battery to show that soda and potash each contained previously unknown metals that reacted violently with oxygen and even with water. They had an extraordinarily high affinity for oxygen. He named these metals sodium and potassium. They were new elements, and Davy was their discoverer. Apart from personal pride, there was national pride involved too, since the Napoleonic Wars were in full swing, and France and England were enemies. Davy's success was even sweeter because it could be seen as a blow struck against French chemistry. Lavoisier had invented the name *oxygen* to identify the gas with acids; oxygen, as we noted before, means "acid producing" or "acid generating," and oxygen for Lavoisier was the acidifying principle. Davy had shown that oxygen was a component of the caustic alkalis, which were the opposite of acidic, and he went on to show that oxygen was also a component of the alkaline earths.

Using the voltaic battery, Davy showed that the alkaline earths, like the caustic alkalis, were compounds containing oxygen and previously unknown metals. His characterization and naming of the alkaline earth metals followed his discovery of them: *barium, strontium, calcium,* and *magnesium* are the names he invented for these metals.* Like the names *sodium* and *potassium,* they are still in use today.

The voltaic battery provided Davy with a tool for analysis. Because it was a tool that seemingly could be made ever more powerful, Davy hoped that it might reveal the true or ultimate elements of bodies, not just the so-called el-

---

*In 1808 Davy first called magnesium *magnium,* since *magnesium* had been used for another metal, which we now call manganese; in 1812, Davy withdrew *magnium* and used *magnesium,* in spite of its former use.

ements that Lavoisier was content with. It all depended on what counted as the last products of analysis. If all that was meant was a list based on the current state of chemical analysis, and if that analysis was almost bound to go further in the future, then the products of analysis were not true elements as far as Davy was concerned. If, on the other hand, the last products of analysis were truly the last such products and no deeper analysis was possible, then chemists would have discovered true elements. How could a chemist know when he had reached the end of analysis? Davy was convinced that building voltaic piles of greater and greater size and power would take him to that end. He was also convinced, like a good Newtonian, that nature was very simple. He believed that this might well mean that there were very few ultimate elements, perhaps even just one.

Instead of reaching this goal, Davy's work with the voltaic pile had led him to the discovery of several new elements, which was frustrating. As if that was not bad enough, another approach that he took to a related problem produced a similarly frustrating result, although in a different way. Lavoisier had said that all acids contained oxygen, and from that rule it followed that the acid produced from sea salt (hydrochloric acid) must contain oxygen. Since analogies are important in chemistry, and Lavoisier's argument here was based on assumed analogies among all acids, Davy began by assuming that Lavoisier was right. He tortured the green gas that he obtained from the acid produced from sea salt, trying to pry its oxygen away from it. He burned diamonds in the gas, using a burning glass to produce the necessary high temperature, but he found no oxides of carbon. He passed the gas between the white-hot arcs of a carbon arc lamp and again failed to detect any oxides of carbon. His consistent and total failure to find oxygen in the green gas, and thus to show the presence of any oxygen in the acid produced from sea salt, was another blow to Lavoisier's theory of acidity. So Davy rejected Lavoisier's name for the supposed constituents of that particular acid, which Lavoisier had called oxymuriatic acid. In 1810, Davy came up with his own name for the green gas obtained from the acid, *chlorine,* which indicates greenness, as it does in chlorophyll, but which says nothing about acid making. The name we use for the acid, hydrochloric acid, was in use within two years of publication of Davy's paper on chlorine.

Davy then noticed an analogy between the salts of hydrochloric acid, that is, chlorides, and those of fluoric salts. He inferred the existence of another substance analogous to chlorine, and in 1813 named it *fluorine.** Because of its extreme reactivity, he was unable to isolate it.

*The name *fluorine* was suggested to Davy by the French chemist and physicist André Marie

In spite of the Napoleonic Wars, Davy was issued a passport for France and French-occupied Europe, and in 1813 he was in Paris, where he met his French rival-colleagues. One of them, André Marie Ampère, behaved more as a colleague than as a rival, and he gave Davy a sample of a dark solid that produced a violet vapor. Davy, who had a traveling laboratory kit in his carriage, rapidly characterized the new substance, recognized that its chemical properties were analogous to those of chlorine, named it *iodine,* and sent a paper on it to the Royal Society of London. Ampère's colleagues in Paris were not pleased that an enemy Englishman had scooped them.

Fluorine (inferred but not isolated by Davy), chlorine, and iodine are three elements in the group that we call halogens. They further extended the list of elements that Davy discovered, and further frustrated his search for underlying simplicity and a very few ultimate elements.

Since Davy did not believe that the elements he kept discovering were ultimate elements, he also did not believe that each element had had its own unique and indivisible atoms. He did not buy Dalton's account of atoms and elements (that was why his award of the Royal Medal to Dalton was such a backhanded compliment). Instead, Davy speculated that there might be just one kind of ultimate atom, a center of forces that might be like Newton's gravitational force. This kind of atomism would make transmutation a theoretical possibility, and Davy did give some thought to transmutation, and even, on occasion, thought that he had come near to achieving it. What were acceptable to Davy were the notions of equivalent rather than atomic weights and Dalton's numerical laws of combining proportions, and he used them in pursuing his very differently motivated researches.

John Dalton's atomic theory and his combining laws were the single most influential package in shaping nineteenth-century chemistry. It was Davy, however, and not Dalton, who gained the greater glory in the Royal Society. But the Royal Society was not the whole world of chemical science. There are very good arguments for identifying Davy's contemporary and rival, the Swedish chemist Berzelius, as the grandfather and perhaps the godfather of nineteenth-century chemistry.

Berzelius: The Uncrowned King of European Chemistry

Jöns Jacob Berzelius (1779–1848) brought together the ideas of Lavoisier with those of Dalton and added his own ideas about electrochemistry. His skill in

---

Ampère, whose name is commemorated by the use of *ampere* as a measurement for electrical currents, just as Volta's name is commemorated by *volt* as a measure of electrical intensity or potential difference.

the laboratory, his enormously influential textbooks, and the energy with which he promoted his ideas made him the most authoritative chemist in the first half of the nineteenth century. He developed the first chemical notation that is recognizably similar to our own, in which letters are used as chemical symbols and the numbers of atoms of each species in a compound are clearly indicated. He applied his electrochemical and atomic ideas systematically, so that composition, reactions, properties, classification, and affinities could all be explained by the same theory. He was the architect of the most successful chemical theory in his lifetime, and it made a truly unified theory of chemistry possible, to a degree not achieved before.

He was Davy's rival in electrochemistry, using the voltaic pile much as Davy did to explore the decomposition of substances. But his concern was not with finding ultimate elements. What he wanted was an understanding of the chemical nature of the atoms of different substances, and he regarded that nature as derived from the electrochemical character of the atoms. Like Davy, he recognized the importance of electronegativity and electropositivity, most clearly revealed in electrolysis, when an electric current was passed through salt solutions. Oxygen was a special element for Berzelius, as it had been for Lavoisier, but in different ways and for a different reason. Berzelius believed that oxygen was the most electronegative element, so he placed it at one end of a table of substances ranked according to the electrical nature of their atoms and corresponding in its order to a table of chemical affinities. The unique electrical character of each element explained its unique chemical behavior, and Berzelius used this insight to classify salts and minerals. This was a matter of great economic significance in a mining country like Sweden, and it may have contributed to Berzelius's rapidly growing reputation in the German states, where mining was also the leading industry.

Atoms had distinct electrical natures, and so did radicals, which were stable groups of atoms that remained together through successive chemical reactions. Every acid contained its characteristic radical. Acid radicals were all electronegative—they all migrated to the positive pole during electrolysis, while metals, being electropositive, migrated to the negative pole. Every chemical compound, according to Berzelius, had an electropositive and an electronegative part, held together by their electrochemical affinities. Here was the foundation of electrochemical dualism in chemistry, and it matched the binary nomenclature that had originated with Linnaeus and been established in chemistry by Lavoisier and his collaborators. Copper sulphate, for example, has positive copper combined with negative sulphate, and the sulphate "radical," as Berzelius called it, was possible because its two constituents (dualism again) differed in the degree of their electronegativity. The course of a chemical reaction depended on how far apart reactants and their constituents were on an electrochemical scale of affinities; similar atoms repelled each other, while unlike ones attracted each other.

Berzelius's use of the atomic theory and of a theory of electrochemistry to account for chemical reactions worked best for salts. The chemistry of salts had been at the heart of much eighteenth-century chemistry, and the work of Berzelius helped to ensure that it remained central in the first half of the nineteenth century. Quantitative analysis, using the balance to establish the weight of each constituent, went along with qualitative analysis, the identification of particular constituents (metals, acids, alkalis, etc.) using the blowpipe and other tests. These analyses, combined with an atomic theory, enabled Berzelius to write formulas for the precise composition of many substances, using chemical symbols that he invented. His formulas, and the symbols on which he based them, were so prominent a part of his influential and widely translated textbooks that they helped to shape the whole of chemical discourse. There were problems, as we shall see later, in establishing these formulas, problems of the kind created by Dalton through the application of *his* simplicity criteria. Still, Berzelius's formulas represented a triumph of analysis and classification. The consistent use of formulas was subsequently incorporated in the writing of chemical equations to show the course of a reaction; chemical equations came into general use around 1835.

Berzelius's methodology had been derived from inorganic chemistry, and especially the chemistry of salts, and it was based on atomism and electrochemical dualism. It was remarkably successful in giving inorganic chemistry coherent shape, with formulas, affinities, and classification firmly in place. But inorganic chemistry, the chemistry that included salts and minerals, was much more easily handled than organic chemistry, the chemistry of those compound substances that exist in nature as constituents of animals or plants, or are derived from such constituents. The next chapter will explore problems and triumphs in the establishment of organic chemistry.

Medicine and pharmacy had traditionally made much use of herbal medicines. In the Renaissance, Paracelsus pioneered the use of mineral substances in the treatment of disease. His reliance on mercury as a cure for syphilis was just one of his many departures from traditional Galenic medicine. Iatrochemistry represented a partnership between chemistry and medicine. That partnership has remained active right up to the present day. Chemistry in the universities was often closely allied to medicine, and Hermann Boerhaave at the University of Leiden in the Netherlands was not only the most influential professor of medicine but also the most influential teacher of chemistry in the early eighteenth century. He and other chemists made particularly close investigations of the chemistry of plant matter. Indeed, Boerhaave devoted the second volume of his two-volume textbook of chemistry to vegetable chemistry. But increasingly, as the eighteenth century progressed, chemical research was more successful where it dealt with mineral acids, with metals and their salts, than where it dealt with vegetable chemistry. We have already learned the importance of the chemistry of mineral salts. In Scandinavia and in German-speaking principalities, the universities, philosophical societies, and mining academies applied and extended the latest discoveries about the mineral kingdom. Theirs was the chemistry of nonliving nature, which became known as *inorganic chemistry.* The chemistry of gases was gradually added to this branch of chemistry and systematized in Lavoisier's chemical revolution.

Following Lavoisier, chemists had a set of rules and a provisional list of elements to work with. Later, using Dalton's atomic theory and his laws of combining proportions, chemists were able to determine atomic weights and to arrive at molecular formulas indicating the nature and number of the atoms in a molecule. Molecules, the smallest part of a compound that possessed the chemical properties of that compound, could then be classified. Berzelius's classification of mineral compounds rested upon his discoveries about the electrochemical properties of atoms, an explanatory notion grafted onto Dalton's simple atomic theory. Chemists were able to establish research programs based on the insights of Lavoisier, Dalton, and Berzelius. For several decades they were, however, far more successful in applying those insights to the inorganic than to the organic realm.

Organic and *inorganic,* as adjectives applied to chemistry, did not come into general use until the 1820s, but they were both used and understood as chemical labels from at least a decade earlier. They were not the words used by chemists around 1800, who referred instead to animal, vegetable, and mineral substances, corresponding to the three kingdoms of nature. Vegetable and animal substances were produced by and found in living bodies, where their activity was related to the processes of life. Some chemists thought it reasonable to explore those processes chemically, whereas others were convinced that living processes were quite different from the nonliving reactions that they carried out in the laboratory. The former group included Lavoisier and Berzelius as well as many chemist-pharmacists and chemist-physicians throughout the eighteenth century. The latter group, believing in a form of vitalism, saw the chemistry of living bodies as beyond the reach of chemical investigation, just as Stahl (Chapter 3) had seen the processes of living bodies as beyond mere chemistry. But even those chemists who regarded the substances produced by living nature, by animals and vegetables, as proper subjects for chemical investigation made less progress in organic chemistry than they did in inorganic or mineral chemistry. That is not to say that they did not make significant progress; they successfully established the foundations for a chemistry of plant substances. But the chemistry based on the work of Lavoisier, Dalton, and Berzelius was more successful—and succeeded earlier—in the inorganic realm.

There are several reasons that organic chemistry at first made slower progress than inorganic chemistry did. Inorganic elements and compounds can often be isolated in a reasonably pure state. Naturally occurring mineral ores may be a source of pure metals, salts, or oxides. In some parts of the world, sulfur exists in a pure state as an element. Atmospheric gases can be separated; other gases, such as hydrogen, can be obtained in reasonably pure form, for example, by the action of a mineral acid on a metal. The *availability of pure substances,* whether naturally occurring or produced in the laboratory, is an important prerequisite for the accurate determination of chemical formulas.

Contrast this availability in the inorganic realm with the substances that are readily available from organic nature. How could chemists extract pure substances from samples of plant tissue or flesh? If plant or animal matter is subjected to distillation, then, typically, a black residue remains in the distillation vessel, and liquid and gaseous substances are driven off. The distillation of different plant substances produces substances that are qualitatively similar, so chemists attempted to distinguish between different plant matters by analyzing the products of distillation quantitatively as well as qualitatively. The solid residue did not vary greatly between different vegetable matters, but there were pronounced differences in the properties and quantities of the liquids yielded by distillation, and these differences were carefully probed. Solvent extraction and fractional crystallization were two of the supporting techniques used to refine organic analyses, but distillation remained the key operation. Varying the conditions of distillation could produce significant differences in the results, and thus eighteenth-century chemists were led to make important advances in standardizing their analytical procedures.

In spite of these advances, major problems remained. The substances produced by distillation were usually complex mixtures, of uncertain purity. There was an added problem: How could chemists know that what they obtained from the destructive distillation of organic matter was present as such in the original matter? Were chemists extracting a substance already present, or did they produce it through their destruction of the original organic tissues? The question was troubling to philosophically sophisticated chemists in the late eighteenth and early nineteenth centuries, as, indeed, it had been to Van Helmont and Boyle in the seventeenth century. Attempts to minimize the destructive violence of analysis by carrying out distillations at the lowest effective temperature did not solve the problem.

Suppose, instead, that chemists fastened on organic substances that were relatively pure, not simply a soup brewed from the whole organism. A typical list from as late as the 1830s identifies as the principal organic substances gelatin, albumen, fibrin, fat, and mucus. More medically oriented texts around the turn of the century would have added chyme and chyle (substances associated with digestive processes) and blood. We know today that these are enormously complicated substances, mostly made up of molecules containing unimaginably large numbers of atoms. How could one begin to determine their molecular formulas, their composition in terms of the kinds, arrangement, and, ideally, the grouping of atoms within the molecule?

There is an added problem. As chemical research was already beginning to make clear before 1800, the principal elements in organic compounds were carbon, hydrogen, oxygen, and nitrogen. All organic compounds turned out to be carbon compounds, for reasons related to carbon's ability to form long chain molecules (a concept well beyond all but the wildest speculation in 1800). Most organic molecules also contain a lot of hydrogen atoms. Hydrogen has the lightest atom of any element. Consider a molecule containing a large number of carbon atoms. It will have a large molecular weight. If, as is likely, the molecule also contains a large number of very light hydrogen atoms, then a small percentage error in analysis will translate into a significant error in determining the number of hydrogen atoms in the molecule. The result will be a formula that is simply wrong. Contrast this with the typical case of an inorganic molecule, which is likely to be made up of a small number of atoms, most of which will be much heavier than hydrogen. A small quantitative error in analysis will be unlikely to lead to an incorrect formula. Thus, inorganic substances are generally easier to analyze and to classify than organic substances are.

For all these reasons, the chemistry of organic compounds, compounds found in or produced by living bodies, was more difficult to make sense of than mineral or inorganic chemistry, and so it was systematized later. There were, however, areas of organic chemistry that were successfully investigated during the eighteenth century, and it is time to look at these, and to see why and how well they worked.

Lavoisier, Again

Lavoisier is mainly remembered for shaping and leading the chemical revolution of the late eighteenth century. His chemistry was variously characterized at the time as the oxygen system, the new pneumatic chemistry, and the antiphlogistic system. But these characterizations reveal only part of the picture. We need to remember that Lavoisier was executed during the Terror following the first stages of the French Revolution and that he died while at the height of his powers, while working on an ambitiously conceived and brutally interrupted research program. That program was not at all limited to the mineral realm.

First, Lavoisier made the chemistry of salts central to his new system of chemistry, but he did not include only those salts formed with mineral acids. He lists a large number of what we would call organic acids and describes their combinations with "salifiable bases," that is, salt formation. The names of many of the acids that he lists are familiar today, including acetic acid, benzoic acid, lactic acid, and oxalic acid, which form acetates, benzoates, and so on. He did not erect a border between these compounds and mineral acids and salts; they were part of a unified chemistry.

Second, Lavoisier was deeply engaged in the study of a discipline that he helped to establish, that of physiological chemistry. In particular, he was interested in understanding the chemistry of respiration. Viewing respiration as he did, as a slow combustion in the lungs, he explored the ways in which respiration and combustion could shed light upon each other. He used human subjects, examining the air that they breathed out and monitoring the air that

Experiments on Respiration in Lavoisier's Laboratory

Here we see Lavoisier carrying out an experiment on human respiration. His experimental subject is the chemist and physiologist Armand Séguin (1767–1835), his assistant from 1789 until 1794. The year 1789 not only witnessed the fall of the Bastille, the prison-fortress symbolizing the old regime, but also the publication of Lavoisier's *Traité élémentaire de chimie* (*Elements of Chemistry*), the manifesto of the chemical revolution. Five years later, Lavoisier was guillotined during the Terror that followed the first phases of the French Revolution.

Research on organic chemistry and physiology was Lavoisier's principal focus during his final years. In this drawing, Séguin is seated at the left with a mask over his face. The experiments examined the rate of conversion of oxygen to carbon dioxide, both when the subject was at rest and when he operated a foot pedal. Respiration was seen as a form of combustion,

providing body heat and energy, and it increased in rate when the body was performing physical work. Lavoisier was seeking to extend his quantification of chemistry into the realm of physiology, an enterprise halted prematurely by his execution.

Mme. Lavoisier drew this sketch, and she has included herself in it, taking notes at a table on the right. She worked closely with her husband, and in the best-known portrait of Lavoisier, by the painter Jacques Louis David, it is Lavoisier who is seated and writing, while Mme. Lavoisier stands with one hand on his shoulder and one on the writing table. Note here the simplicity of the apparatus, with its pneumatic trough, bell jar, glass globe, and funnel, a far cry from the elaborate and sophisticated gasometer that Lavoisier described in his *Traité*, but typical of much of his research apparatus.

■ Sketch by Mme. Lavoisier, first published in E. Grimaux, *Lavoisier 1743–1794* (Paris, 1889), opposite p. 128.

they breathed in. He also used nonhuman animal subjects and measured the heat they produced in respiration, just as he had measured the heat produced in combustion. The ice calorimeter that he and Laplace had designed was quite big enough to study the respiration of a guinea pig over an extended period of time.

Heat balance, based for Lavoisier upon the conservation of caloric, was something that could be studied in mineral chemistry through heats of combustion and specific heats, and in animal chemistry through heats of respiration. Heat was very much part of Lavoisier's chemistry, as, in differing and more limited fashion, it had been for Priestley and phlogistic chemists before him. Monitoring the products of respiration, carbon dioxide and water, contributed to Lavoisier's understanding of the chemistry of gases. He was aware of the symmetry between animal and vegetable respiration, with animals converting oxygen to carbon dioxide and plants principally converting carbon dioxide to oxygen. This effective symbiosis, in which animals and plants contributed to one another's vital processes, each through restoring to the atmosphere a gas needed for the other's respiration, was not Lavoisier's discovery. His interpretation was, however, his own, and it was novel. It further reinforced his underlying methodological assumption, which came to have the force of conviction, that chemistry was a single discipline, encompassing the three kingdoms of nature, animal, vegetable, and mineral. Methodologically, there was only one chemistry.

That meant that Lavoisier's quantitative method, resting upon the tacit principle of conservation of mass, was a key tool in his exploration of the organic as well as the inorganic realm. By 1785, Lavoisier had determined the proportions of the constituents of carbon dioxide (Joseph Black's fixed air) and also the quantitative composition of water. In the next few years, years in which the *Essay on Chemical Nomenclature* (1787) appeared and which culminated in his *Elements of Chemistry* (1789), Lavoisier concentrated his researches on the plant kingdom. He tackled three main problems: fermentation, the germination and growth of vegetables, and the composition of plant matters. In each case, he used the quantitative methods of the balance sheet, based upon conservation, to arrive at a picture of the transformations of matter. In studying plant composition, he used traditional methods such as distillation, and he also relied extensively on an analysis of the combustion products of various oils and other vegetable matter. The oxygen theory of combustion, the composition of water, and quantitative analysis came together just as they did in mineral chemistry.

The word *fermentation* had for centuries covered a wide range of chemical processes, all of which involved a degree of effervescence, the production of heat, and a change in properties. We came across Paracelsus's use of the term to indicate any process involving chemical change, growth, or development, with Nature as God's alchemist, the initiator of fermentation. There was also an alchemical usage, in which one substance would act as ferment, such as yeast in brewing and baking; the philosophers' stone was the supreme seed or source of fermentation. The rise of sap in plants and the germination and growth of plant seeds were other examples of fermentation. Lavoisier, however, developed a much more tightly focused view of fermentation, and the narrowing of the term that followed was partly the result of his work.

At one time, Lavoisier interpreted the composition of plants in terms of water (or hydrogen and oxygen) and carbon. He came to realize that these substances were incorporated into more complex ones. Gradually his chemical studies of what he described as the decomposition of water by plants showed clearly that there were sugars in plants that decomposed during fermentation and were converted to water and fixed air. That, and other work, led him to a more direct study of the fermentation of sugar solutions by the addition of yeast and to the development of apparatus specifically designed to measure the products of that fermentation. Here he was studying what came to be understood as the central process of fermentation, although his use of the term was never quite as narrow as ours. He combined gravimetric, volumetric, and qualitative analysis with the design and use of instruments very similar to those devised for studying gas chemistry in the inorganic realm. For Lavoisier, there was only one chemistry.

Berzelius and Laurent: A Question of Method

There was also only one chemistry, methodologically speaking, for Berzelius, who in 1814 published a series of quantitative analyses of some relatively simple organic substances. He explained that he had undertaken the analyses in order to determine how far organic bodies obeyed the laws governing inorganic ones. He carried out the analyses with great care and precision (the work took him years before he was satisfied, in part owing to problems in obtaining sufficiently pure samples), presented them in terms of the numbers of atoms of oxygen, hydrogen, and carbon, and published the results with considerable satisfaction. It was possible and fruitful, he concluded, to explore the realm of organic chemistry using the methods developed in inorganic chemistry, since the two realms obeyed the same chemical laws. These included Dalton's laws of combining proportions.

For Berzelius, the single most important law, the one that owed most to his own researches and that distinguished his approach, was that of electrochemical dualism. According to Berzelius, atoms and radicals combined most vigorously with those of opposite electrical character, so his scheme could be characterized as a dualistic one. It was electrical character that above all else determined the behavior of atoms and radicals (such as sulphate and nitrate radicals), and thereby also determined the constitution and properties of the compounds they formed. With this guiding law, Berzelius was able to interpret reactions, explain the properties of compounds, and classify them. Not surprisingly, since Berzelius had arrived at the law by exploring the formation and behavior of mineral salts, the law worked well for such salts. Berzelius had then asked whether the same law, and the same kinds of interpretation of results, would work first across the mineral kingdom, and then in plant chemistry. His answer was a firm yes. Berzelius's results, and the publicity his writings gave them, earned him a position of authority in European chemistry.

By the late 1830s, Berzelius was the leading chemist of Europe, and his views dominated European chemistry, especially in the German states (Germany was not yet a unified nation). Berzelius's chemistry, for all its importance and innovation, was in one key respect conservative. He believed that chemists should proceed from the known to the unknown and unless the evidence told them otherwise, they should use tried and true methods in exploring new fields. That meant using the methods of mineral or inorganic chemistry in exploring organic chemistry. Berzelius's first published organic analyses persuaded him that these were the right methods to use.

In 1836 a young, obscure French chemist named Auguste Laurent (1807–53) wrote a doctoral thesis presenting a new theory of organic combination and criticizing electrochemical dualism. Laurent argued that chemical formulas should express the totality of reactions undergone by a compound. But the same compound would behave differently with different reagents and under different experimental conditions. Berzelius's formulas did not allow for this diversity. Laurent suggested that a way out of this difficulty lay in thinking of chemical reactions as determined principally by the way atoms were arranged in a molecule. Laurent had evidence to support this view. Acetic acid, a compound containing a methyl radical (CH_3) combined to a carboxyl acid radical (COOH), with a formula that we write today as CH_3COOH, changed into chloracetic acid, $CClH_2COOH$, or even into trichloracetic acid, CCl_3COOH, when chlorine gas was bubbled through it. To some chemists this was a case of *substitution,* with chlorine substituting for hydrogen, that is, replacing it and playing its part. Chloracetic acid had properties very similar to those of acetic acid, but chlorine was a heavy electronegative atom and hy-

drogen was a light electropositive atom. How could an electronegative atom
take the place of an electropositive one? How could the presence of such different atoms not lead to major differences in the chemical properties of these
two compounds? This could only be the case if the geometric arrangement of
atoms in the molecule was more important than the nature of the individual
atoms.

So, at least, was Laurent to argue. He wrote to Berzelius, urging him to recognize the virtues of an explanatory scheme based upon the idea that similar
arrangements of atoms in a molecule were responsible for similar properties in
compounds, while different arrangements of the same atoms produced different properties. Laurent argued that there were examples in inorganic chemistry
where this was clearly true. For example, there were different forms of sulfur
(one was crystalline; another, an amorphous powder; and yet another, a rub-
bery substance). These forms all contained sulfur atoms, but they were different from one another. This was called *allotropy*, and to Laurent it argued for an
explanation based upon the arrangement of atoms. Then there were cases
where different minerals could form mixed crystals, all having pretty much the
same shapes although containing different kinds of atoms. This was called *isomorphism*, and again for Laurent it argued in favor of a theory that derived
properties from the arrangement of atoms in a molecule. Isomorphism and allotropy were inorganic analogs of a phenomenon first discovered in organic
chemistry, where two molecules containing the same number and the same
kind of atoms could differ greatly in properties. For example, urea and ammonium cyanate have the same composition but different properties. Several
chemists had discovered this phenomenon independently in the first decades
of the nineteenth century, and it was Berzelius himself who coined the term
isomerism for it.

Laurent made the case that isomerism, isomorphism, and allotropy all indicated that the properties of molecules depended more on the arrangement
than on the nature of their constituent atoms. He found illustrations of such
phenomena throughout the realm of organic chemistry, although he was also
drawing on his knowledge of crystallography, itself traditionally related to
mineralogy and inorganic chemistry. But Laurent stressed the organic analogy;
indeed, he insisted that he had derived his own explanatory scheme, based
principally upon arrangement, from the realm of organic chemistry, and then
applied that scheme to a variety of inorganic phenomena. He argued that organic chemistry should provide the unifying model for the whole of chemistry.
Where Berzelius argued from inorganic to organic chemistry, Laurent argued
in the reverse direction. He engaged Berzelius in a debate through a series of

letters on this subject. Berzelius was the leading chemist of Europe, and Laurent was a very junior scholar. Berzelius was not at all persuaded by Laurent's arguments, and he simply told Laurent that they would have to agree to disagree and to see where their different approaches would lead.

Laurent and Gerhardt: What Can We Know, and What Can We Do?

Laurent's work was at first generally ignored, and he remained an obscure professor in Bordeaux. In 1845 he began to work with another provincial chemist, Charles Gerhardt (1816–56), who came to Paris three years later. They were in complete agreement that organic chemistry was the most promising area for chemical research and that it would provide the model for future research across the entire spectrum of chemistry. Organic chemistry would show inorganic chemists the way to advance. Berzelius and his electrochemical dualism were obstacles to progress.

Laurent admitted that he did not know anything about the actual arrangement of atoms in a molecule. For Gerhardt, this meant that he should say nothing about arrangement. But Laurent was convinced that chemistry advanced in the light of ideas as well as by experiment, and his guiding idea was that compounds with similar chemical properties were likely to have similar arrangements of their constituent atoms. Laurent and Gerhardt differed about the usefulness of hypotheses, and they admitted that they did not know and indeed might never know anything about the actual arrangement of atoms in molecules and about the paths taken by these atoms during chemical reactions. But this ignorance, this present and perhaps perpetual lack of knowledge, did not rule out new theories and progress in chemistry. Both Laurent and Gerhardt were trying to bring order to the increasing complexity and chaos of organic chemistry. There were so many different substances, with new ones being discovered almost daily, and it was urgently necessary to devise some fruitful scheme for classifying these substances.

Laurent and Gerhardt agreed that what they could do was to classify organic compounds and to reveal similarities and regularities among them. Analogy was for both of them the key to insight and progress. Laurent believed that substances with analogous properties had analogous formulas deriving from unknown but analogous arrangements of atoms. Classes of compounds (e.g., alcohols, ethers, and salts) would have similar formulas, and Laurent found it helpful to think of these similarities as arising from similar although unknown arrangements of atoms. Gerhardt argued that since the arrangements were unknown, he and Laurent should restrict themselves to the principle that analogous formulas corresponded to analogous properties. And, in the end, that is what they did.

Two very productive sets of ideas emerged from their collaboration, both of which are generally associated more with Gerhardt than with Laurent. Gerhardt published the ideas, but they arose from a true collaboration. The first idea was that of a *homologous series*. To take one of the simplest cases, consider the paraffins. These form a series of compounds composed exclusively of carbon and hydrogen. Each member of the series has, broadly speaking, similar chemical properties, including combustion to yield water and carbon dioxide. The members of the series can be arranged in a sequence of increasing molecular weights, which is also the sequence of increasing density, from gaseous methane through liquid paraffins to solid ones. Gerhardt described such a series as a homologous one, and he asserted that constructing analogous formulas for the series members could represent the analogies of physical and chemical properties. The paraffin series can be represented by the sequence CH_4 (methane), C_2H_6 (ethane), C_3H_8 (propane), and so on, which can be generalized in the formula C_nH_{2n+2}. Similarly, the alcohols (methyl, ethyl, propyl alcohol, etc.) can be represented by the formula $C_nH_{2n+1}OH$. Such series, represented by formulas that increase according to a simple arithmetic progression, are homologous. They make possible a wonderfully increased economy in classifying organic compounds and, through the use of homologous series such as paraffins and alcohols, a corresponding economy in learning organic chemistry.

The other key idea that we owe to Gerhardt and to his collaboration with Laurent is that of *types,* which Gerhardt proposed in 1853, the year of Laurent's early death. The theory of types was based on the idea of substitution. Gerhardt proposed that there were three fundamental types and that all organic molecules could be shown to correspond to one of the types. All organic compounds could be envisaged as derived from three simple inorganic molecules (hydrogen, with two hydrogen atoms; water, with two hydrogen atoms associated with one oxygen atom; and ammonia, with three hydrogen atoms associated with one nitrogen atom). Gerhardt later divided the hydrogen type into two, corresponding to H—H and H—Cl, with like and unlike atoms respectively. Various other atoms or groups of atoms could be substituted for one or more of the original hydrogen atoms. Thus ethyl alcohol (C_2H_5OH) could be seen as derived from the water type by the substitution of an ethyl radical (C_2H_5) for a hydrogen atom in water. Now substances could be placed in their appropriate homologous series and also classified in terms of their type. Type rested on substitution, a concept invented within organic chemistry; but the three type molecules, hydrogen, water, and ammonia, were simple inorganic molecules. Thus type theory offered a bridge joining organic and inorganic chemistry.

Liebig and Wöhler: Radicals and Conservatives

Laurent and Gerhardt were not the only chemists trying to bring order out of chaos. They had attacked Berzelius's electrochemical dualism. Laurent, in particular, by arguing that the arrangement of atoms in a molecule counted for more than the nature of individual atoms did in determining the properties and reactions of a compound substance, had confronted Berzelius and his followers head on. Substitution (for example, of chlorine for hydrogen in acetic acid to form monochloracetic acid, or even trichloracetic acid) presented electrochemical dualists with a real difficulty. If one accepted that true substitution had occurred, with chlorine taking the place and playing the part of hydrogen in the acid, then dualism was in deep trouble. Clearly, according to electrochemical dualists, the interpretation of the phenomenon and even the name *substitution* was wrong.

Instead of exploring and interpreting the whole of chemistry from the organic end, as Laurent and Gerhardt had done, Berzelius's followers agreed that inorganic chemistry was still the way to go. Both sides in the debate asserted that there was ultimately only a single chemistry, and the question became one of method and effectiveness. Chemists knew much more about inorganic compounds than they did about organic ones. Berzelius's followers argued that it was only common sense to move from the known to the unknown. That meant using inorganic chemistry as the secure methodological foundation for the investigation of the organic realm.

What did this mean in practice? First, it meant hanging on to electrochemical dualism as the key to formulas and reactions. It also meant bringing the idea of *preformation* up-to-date. Chemistry had long been concerned with the issue of preformation, with those groups of atoms or elements that persevered through a series of reactions. In Lavoisier's work, dualism (although his dualism was not electrochemical and was not based on Dalton's atomic theory) combined with preformation to make the concept of radicals a central one. There were, according to Lavoisier, acid radicals that existed in acids and were also found in the salts produced by the reaction of those acids with bases (alkalis). The acid radicals existed before the salts were produced; the radicals were *preformed*. The reaction of an acid with a base, and the composition of salts from a metal and an acid radical, were examples of dualism. Inorganic radicals, then, were an essential part of inorganic chemistry. Dalton's atomic theory and Berzelius's electrochemical dualism had refined the picture, so that radicals became understood as stable and persevering groups of atoms that functioned as units, that had particular electrochemical properties, and that were in many ways analogous to atoms. The sulphate radical in sodium sulphate, for example, functioned analogously to chlorine in sodium chloride.

The question then became, for those who sought progress in organic chemistry using models derived from inorganic chemistry, Are there organic radicals, analogous to inorganic radicals? The first organic radical was identified by two German chemists, Justus Liebig (1803–73) and Friedrich Wöhler (1800–1882). They were both deeply influenced by Berzelius and had been looking for ways to navigate the dark and unknown forests of organic chemistry using Berzelius's theories as their guide. In 1832, long before type theory and homologous series existed, they published a paper on their discovery of the benzoyl radical, a substance that persevered through a variety of compounds, including benzaldehyde, benzoyl chloride, and benzoic acid. What is significant about their paper is that an idea that had been enormously fruitful in inorganic, or mineral, chemistry had been deliberately used to bring order to a region of organic chemistry. The discovery of other radicals followed.

The type theory of Gerhardt and the radical theory of Liebig and Wöhler both contributed to the unification of chemistry. In spite of the opposition of radical and type theory, both underwent constant modification over the ensuing decades. Type and substitution theorists began to admit that it was perhaps necessary to give more weight to the properties of individual atoms than they had formerly believed. Radical theorists gradually agreed that the position of atoms in a molecule was important. As one prominent nineteenth-century chemist observed, looking back on this process, "the two theories, the dualistic radical theory and the unitary substitution theory, were both true and imperfect, ... They underwent gradual development, scarcely influenced by each other, until they have come to be almost identical in reference to points where they at one time seemed most opposed."*

*Alexander Crum Brown, *Report of the British Association for the Advancement of Science* (1874): 46, 49.

Lavoisier had provided the rule for building a table of elements. If chemists could decompose a substance, then that substance was a compound. If they could not decompose it, then it was to be regarded as an element. Its elementary status was, however, provisional and open to challenge, since what chemists could not decompose today, they might well be able to decompose tomorrow. In that case the substance would lose its status as an element and be revealed as a compound.

Then came John Dalton, with his atomic theory. *Atom* is derived from the Greek word meaning "indivisible." Dalton associated his atoms with elements, each element being characterized by its own unique atoms, with unique properties, including weight, *atomic* weight. For those who fully accepted Dalton's ideas about atoms and elements, it was hard to think of elements as having merely provisional status. It was tempting to think of them as consisting of true indivisible atoms, which implied that they were ultimately unchanging, except in the combinations that they formed with other atoms. Determining the atomic weights of elements thus meant finding something enduring, and the research programs devoted to determining atomic weights were, for the most part, confident about this aspect of atoms and their weights. In the light of these programs, molecular weights and formulas indicating the atomic composition of compounds also promised to be enduring.

Chemists, like other kinds of natural philosophers, were also attracted by ideas about the regularity and ultimate simplicity of nature. We have seen that Dalton's atoms were individually simple, although there were a lot of different kinds. The list of elements stood at around thirty when Dalton published his atomic hypotheses or theories. It seemed to some chemists that simplicity and regularity might well result in a set of atomic weights that were all whole numbers.

Keep It Tidy, Keep It Simple: Prout, Gay-Lussac, and Avogadro

This was the view of atomic weights advocated by the English chemist and physiologist William Prout (1785–1850), but he combined it with other ideas that undermined the idea of indivisible atoms. Prout regarded hydrogen as the "primary" material agent in mineral chemistry, and he believed that it was converted, by galvanic electricity and other means, into all the other chemical elements. There was an original or prime matter, from which all the elements, and thus all bodies, were composed. Heavier elements were built up of multiples of the fundamental unit, the hydrogen atom. It followed that, if the atomic weight of hydrogen was taken as one, the atomic weights of the heavier elements would be integral multiples of that unit. In other words, atomic weights would all be whole numbers.

It is a striking fact that many atomic weights, although not all, *are* close to whole numbers, and the distribution of atomic weights is in general so close to whole numbers that there has to be a reason for it. We know today that the reason is that the heaviest part of any atom is its nucleus, made up of neutrons (uncharged particles) and protons (positively charged particles), which have almost exactly the same weight. Some elements have an atomic weight that is not a whole number, for example, chlorine has an atomic weight close to 35.5. From the discovery of atomic structure around the beginning of the twentieth century, we now know that such elements have two or more isotopes. Different isotopes of a given element are forms of that element having different numbers of neutrons in the nucleus of their atoms, and therefore have different atomic weights, since the atomic weight of an isotope is the sum of the number of protons and neutrons in the nucleus of its atoms.* Atomic weights that are not whole numbers arise from the mixture of isotopes, whose weights average out to a non-integral value. All that Prout knew or could know, however, was that many atomic weights are either whole numbers or very close to whole numbers. In two papers published in 1815 and 1816, he put forward what has become known as *Prout's hypothesis,* that atomic weights are all whole numbers because they are multiples of the weight of the hydrogen atom.

Chemists with more confidence in the regularity and simplicity of nature than in the results of rigorous and painstaking analyses were able to round off even problematic atomic weights to whole numbers, without doing too much violence to the results. Molecular weights could then be calculated from the results of quantitative analysis. The Scottish chemist Thomas Thomson (1773–1852), who had been an early convert to Dalton's atomic theory, publishing an account of it even before Dalton himself, promptly set about showing that

*Electrons, negatively charged particles, are also constituents of matter and contribute to atomic weights. But since electrons weigh only about 1/1,840 of the weight of a proton, they may be ignored for all but the most precise determinations of atomic weights. The sum of the weights of protons and neutrons is close enough.

Prout was right. If you know what the result should be, it is relatively easy to obtain it. Thomson accordingly came up with a set of atomic weights supporting Prout's hypothesis. He was promptly criticized for the unsatisfactory nature of his analyses. Berzelius, a scrupulously careful chemist, was ruthlessly dismissive of Thomson's results. The best that one can say of this aspect of Thomson's work is that he was careless. Prout's hypothesis had received a setback, but it had a way of coming back to life throughout the century, like a phoenix.

The same could be said about Dalton's atomic theory, but that theory, unlike Prout's, did, in modified form, become a fixture in chemistry. Dalton's ideas had led to a system of atomic weights and also to some rules or laws of combining proportions, including the law of multiple proportions.* Eight parts by weight of oxygen combine with six parts by weight of carbon in forming carbon monoxide; in carbon dioxide, twelve parts by weight of carbon combine with the same eight parts by weight of oxygen. The ratio of the weight of carbon in the dioxide to its weight in the monoxide is $12:6 = 2:1$.

There is of course a problem, which did not arise for Dalton. His simplicity criterion about the ratios of combining *atoms* works for carbon monoxide and carbon dioxide, but it denies the possibility of diatomic molecules such as H_2 and O_2. This leads to problems, and it produces headaches when applied to water and hydrogen peroxide, H_2O and H_2O_2 for us, but HO and HO_2 for Dalton. Errors in atomic weights lead to errors in determining molecular formulas, and errors in molecular formulas lead to errors in determining atomic weights. Dalton erred by a factor of two in the atomic weights of oxygen (8, instead of our 16) and carbon (6, instead of our 12). These errors gave rise to confusion and controversy in the ensuing decades.

Dalton wrote down his ideas about atoms in notebooks of 1802. The first public mention of Dalton's atomic theory and laws of combining proportions was by Thomas Thomson in 1807, and only in the following year, 1808, did Dalton publish his own account. That publication coincided with the publication in France of a different law of combining proportions.

Joseph Louis Gay-Lussac (1778–1850) was one of the brilliant generation of French chemists and physicists who came to prominence after the French Revolution. He made balloon ascents to study the earth's magnetism and to collect samples of air. In 1805 he published a paper showing that oxygen and hydrogen formed water by combining in the ratio of 1:2 by volume. Other gas

*Dalton's law of multiple proportions states that when two elements combine in a series of compounds, the ratios of the weights of one element that combine with a fixed weight of the second are in a ratio of small whole numbers (see Chapter 7).

studies soon showed him that there was a *law of combining volumes* for gases that corresponded to the law of multiple proportions for analysis by weight (gravimetric analysis). When gases combined, they did so in volumes that were in a ratio of small whole numbers. Gay-Lussac's gas studies convinced him that regularities in gas composition were more general and simpler than those regularities that Dalton discovered for combining weights. Dalton and Gay-Lussac, the one wedded to gravimetric analysis, the other to volumetric analysis, did not develop any enthusiasm for each other's work, and it was by no means obvious how their laws of combining weights and combining volumes could be reconciled. Neither for the first nor for the last time in the history of chemistry, French and English chemists were rivals when they might better have collaborated with one another. Intellectual pride was reinforced by the hostilities of the Napoleonic Wars.

There was a way in which Dalton's law of multiple proportions and Gay-Lussac's law of combining volumes could be reconciled, but even when it was offered, it was at first not well received. In 1811, the Italian chemist Amedeo Avogadro (1776–1856) proposed an explanation for the regularities described in Gay-Lussac's law. He suggested, in what has become known as *Avogadro's hypothesis,* that equal volumes of gases at the same temperature and pressure contain equal numbers of particles. These particles could be atoms or molecules, simple or compound. They were the smallest particles of oxygen gas, hydrogen gas, carbon monoxide, carbon dioxide, ammonia vapor, and so on. A consequence of Avogadro's hypothesis was that combination by volumes in the ratio of small whole numbers implied the combination by particles in the ratio of small whole numbers. This was a remarkable insight that brought together two empirical laws, Dalton's combining proportions and Gay-Lussac's combining volumes, and one might have expected that it would have been welcomed by the whole chemical community of Europe.

Such an expectation would have been wrong. Avogadro's hypothesis did not find a general welcome and acceptance for a half century. Why not? Let us begin by considering Gay-Lussac's own results for the composition of water. Two volumes of hydrogen combine with one volume of oxygen to give two volumes of water vapor. That implies, according to Avogadro, that two particles of hydrogen combine with one particle of oxygen to give two particles of water vapor. That can happen only if each particle of oxygen is divisible into two parts, as it is in our modern formula for oxygen gas, O_2. But for Dalton and his followers, diatomic molecules of a single element cannot exist. He was convinced that identical atoms do not and cannot combine directly with one another, as only unlike atoms could combine. And if oxygen was not diatomic, but monatomic, then Avogadro's hypothesis implied that each atom of oxygen was divisible into two parts, another impossibility for Dalton, since his atoms were by definition indivisible. Avogadro's hypothesis was simply not compatible with the assumptions of Dalton's atomic theory.

Dalton's was not the only atomic theory at the time. Berzelius had also proposed an atomic theory, but with divisible atoms. Berzelius, however, was also unhappy with Avogadro's hypothesis. Divisible atoms might be all very well, but diatomic molecules containing two identical atoms violated Berzelius's developing ideas about the electrical nature of atoms and the relationship between electrical character and chemical affinity. Substances that had affinity for one another, and were therefore disposed to combine together, had to have opposite or at least significantly different electrical characters, as was the case with electropositive hydrogen and electronegative oxygen or electropositive sodium and electronegative oxygen. By allowing for the possibility of diatomic molecules of oxygen, hydrogen, and other gases, Avogadro was making claims that ran counter to Berzelius's electrochemical dualism.

There were other, less theoretical but no less persuasive objections. Some substances, such as ammonium chloride, dissociate in the vapor phase. That is, a single particle of vapor turns into two or more particles. Two or more particles occupy two or more times the volume that one particle does. That wreaks havoc with measurements of gas volumes and provides empirical evidence that fails to obey Gay-Lussac's law, making apparent nonsense of Avogadro's hypothesis. It was not until the phenomenon of dissociation was understood, and interpreted in terms of reaction kinetics, that this objection could be countered. Similar objections were raised against Dalton's laws of combining proportions, which work only for compounds of fixed composition. Metallic alloys and salt solutions, to take two of the most obvious exceptions, do seem to share some of the characteristics of chemical compounds, but they do not fit Dalton's laws. The simplest way to avoid that objection was to say that only those substances that did fit Dalton's laws were true chemical compounds, but that is a circular argument that did not convince critics.

In the decades when Avogadro's hypothesis was mostly gathering dust, Berzelius's electrochemical dualism was the most successful chemical theory in Europe. It explained the properties of elements and compounds, especially the chemical reactions they underwent, and Berzelius, looking at those properties and reactions, derived his own formulas for compounds. But, as Berzelius's critics from organic chemistry pointed out in the 1840s, different reactions suggest different formulas for one and the same substance, and that is not very satisfactory. What is more, the kinds of problem arising from Dalton's simplicity criteria, leading many chemists to err by a factor of two in determining a variety of atomic weights, meant that there would continue to be major disagreements about fundamentals, including atomic weights, molecular weights, and molecular formulas. Prior to 1860, chemists who avoided these problems were in a minority.

Among that minority, Gerhardt and Laurent, thorns in Berzelius's side in the final decade of his long life, succeeded in confronting and resolving many of the problems concerning atomic and molecular weights. The route they took was not widely shared, but it did involve them in adopting a rule corresponding to Avogadro's hypothesis. Once again, they were in the minority at the time, but their molecular formulas tended to have the right number of atoms, two in a hydrogen molecule, two in oxygen, three in water, and so on.

They were in a minority because most chemists at the time derived atomic weights from relative combining *weights* and equivalents. Laurent and especially Gerhardt looked instead to combining gas *volumes* and to a whole group of various other properties, including the physical property of specific heats.* Those who worked from combining weights used simplicity rules, including the main one that atoms combined with one another in the simplest proportions by weight and number. This was a fruitful hypothesis, but it was just a hypothesis, and so, as Gerhardt and Laurent pointed out, atomic weights based on it were also hypothetical. They also pointed out that the use of proportional numbers combined with simplicity rules often led to fractions of atoms, as for example in ferrous oxide, where the proportional numbers require two-thirds of an atom of iron to one of oxygen. This was not very satisfactory. Of course, one could always triple the numbers and arrive at two atoms of iron to three of oxygen, but that seemed arbitrary.

Analogy had always been a guide to theory and practice in chemistry. It was therefore reasonable to assume that elements with similar specific heats had similar atomic weights. Two other French scientists, Pierre Louis Dulong (1785–1838) and the short-lived Alexis Thérèse Petit (1791–1820), explored the relation between atomic weights and specific heats. They found that, for most elements, the product of these two quantities (the "atomic heat" of elements) was approximately a constant. This, known as *Dulong and Petit's law,* was published in 1819. But the law held only if chemists doubled some of the atomic weights arrived at by the use of Dalton's rules. Dalton, naturally, was not pleased with this suggestion, and few chemists took any notice of what they regarded as an essentially physical rather than chemical result. Disciplinary

---

*The specific heat of a substance may be defined as the amount of heat required to raise one gram of the substance by one degree Celsius.

boundaries can keep ideas out as well as in. Laurent and Gerhardt, however, working with chemical and physical analogies in arriving at their formulas, made good use of the specific heat data and of analogies in physical as well as chemical properties in determining atomic weights and molecular formulas.

There were problems with specific heats as a guide to atomic weight. Some elements, such as sulfur, existed in different allotropic forms having different specific heats. Did this mean that such an element had more than one atomic weight? Again, volumetric work on vapors revealed that different allotropes of sulfur occupied different volumes. In such cases, even Laurent and Gerhardt had to fall back on proportional weights to arrive at atomic weights.

The fundamental rule that Gerhardt and Laurent adopted was Avogadro's volume hypothesis, which stated that equal volumes of gases contained equal numbers of particles. In order to make sense of this hypothesis, Laurent proposed that "each molecule of an element can be divided into two or more parts that we call *atoms;* these molecules can be divided only in the case of chemical combinations."* So, for example, molecules of gaseous hydrogen and oxygen contained two atoms apiece. When they combined, experiment showed that one volume of oxygen combined with two volumes of hydrogen to form two volumes of water vapor, and this could be represented by the equation $2H_2 + O_2 = 2H_2O$. Gerhardt and Laurent used superscripts rather than subscripts, so that they represented a diatomic molecule of oxygen as O^2. But their formulas, corresponding to the formation of two volumes of water in the above equation and conforming to Avogadro's hypothesis, were otherwise more often than not the same as the ones we use today. They called them "two volume" formulas, and Gerhardt in particular used the words *volume* and *atom* to represent the same thing.† This can be confusing for us today.

In practice, Gerhardt, and Laurent with him, agreed with Berzelius's atomic weights for many bodies, but, for example, Gerhardt's use of the two-volume hypothesis led him to adopt atomic weights for the metals that were half the generally accepted atomic weights. The result was a set of molecular formulas that were simpler than Berzelius's, that brought out physical and chemical analogies between different compounds, and that did not depend on any par-

*Quoted in W. H. Brock, *The Norton History of Chemistry* (New York: Norton, 1993), 229. The Italian chemist Stanislao Cannizzaro proved in 1858 that Laurent was mistaken in thinking that all elements had diatomic molecules. Gaseous nitrogen, hydrogen, chlorine, and oxygen are among those elements whose molecules do contain two atoms apiece, but most nongaseous elements do not have diatomic molecules.

†One molecule of oxygen, O_2, corresponds to one measured volume of gas, in accordance with Avogadro's hypothesis, but for Gerhardt the molecule represented two "volumes," meaning two atomic weights' worth of oxygen.

ticular or privileged chemical reaction, where different reactions could lead to incompatible formulas for the same substance. For the first time ever, the mineral acids and their metallic salts were given molecular formulas generally agreeing with ours today. But, as we have seen, Dalton, Berzelius, and following them many other chemists not only lacked the benefit of our hindsight but had good theoretical reasons for dismissing Avogadro's hypothesis. Because Gerhardt and Laurent proposed formulas based on the acceptance of that hypothesis, their formulas were at first rejected by most chemists. And besides, Berzelius's authority in particular remained strong through the 1840s, and it was not until well into the next decade, after Berzelius had died in 1848, that Gerhardt's and Laurent's ideas met with cautious acceptance. Avogadro's hypothesis had to wait even longer before it was accepted.

Confusion and Resolution of Atomic Weights: Cannizzaro Sets a Standard

Failure to agree whether the atomic weights of metals should be half of what Berzelius advocated was one of the reasons there was little agreement about what formulas, and even what kind of formulas, should represent molecular composition. But there were other issues that led to lively debate about atomic weights. These issues arose from reflecting about some remarkable numerical patterns and regularities in atomic weights, from thinking about allotropy, and from the steady growth in precision in quantitative analysis.

That there were striking regularities in atomic weights had been known for some time. We have seen that Prout's original hypothesis had been prompted by the fact that most atomic weights appeared to be either whole numbers or very nearly whole numbers, and so might be built up of units corresponding in weight to the hydrogen atom. It surely could not be merely by chance that so many atomic weights were close to integers. Thomas Thomson's careless but attractive analyses had reinforced this view. But there were simply too many atomic weights that were not whole numbers, and Prout's hypothesis failed to win acceptance by most chemists. Subsequently, Prout was to contemplate the notion that there might be building blocks smaller than the hydrogen atom, corresponding to perhaps a half, perhaps a quarter of that atom.

Meanwhile the German chemist Johann Wolfgang Döbereiner (1780–1849) observed, between 1817 and 1819, that there were several sets of groups of three elements (triads) where the atomic weight of one was very nearly halfway between the atomic weights of the other two. These triads were of recognizably similar elements, including the halogens (chlorine, bromine, and iodine) and the alkaline earth metals (calcium, strontium, and barium). Döbereiner also identified the triad of sulfur, selenium, and tellurium, and later added the alkali metals lithium, sodium, and potassium. Elements within such families were related in a way that involved regular increases of atomic weight from one element to the next.

These regularities might suggest, as they did to some chemists, that atoms were made of building blocks, one element in a triad differing from the next by the possession or lack of a certain number of such building blocks. Something like Prout's hypothesis might after all prove to be on the right track. Allotropy also suggested that atoms might be compounds, but somehow compounds of identical building blocks, a difficult and unfamiliar concept when Laurent and others first proposed it. The existence of triads, the existence of families of elements, and the phenomenon of allotropy all contributed to a revival of Prout's hypothesis. But since atomic weight determinations had clearly shown that any building block had to be smaller than an atom of hydrogen (how else could one explain atomic weights that were not whole numbers?), the original hypothesis clearly would not do. Instead, the revival was based on smaller units, fractions of the atomic weight of hydrogen. Going back to one of the most obvious problems for the original version of Prout's hypothesis, chlorine had an atomic weight of 35.5. If the subatomic building blocks of prime matter corresponded to half an atom of hydrogen, then chlorine's atomic weight would not be a problem after all. Prout himself had thought along these lines.

Berzelius, who died in 1848, wrote a critique of Prout's hypothesis in 1845. He noted that transmutation had never been observed in the laboratory. If Prout's hypothesis was right, transmutation would be at least a theoretical possibility, and failure to observe it argued against Prout. The experimental data about atomic weights that best supported Prout's hypothesis were those of Thomas Thomson, and Berzelius had nothing but scorn for Thomson's abilities as an analyst. Clearly, Berzelius regarded it as merely a coincidence that many atomic weights had either whole-number or half-whole-number values.

There were, however, increasingly accurate atomic weights, determined with great precision by laboratory chemists whose practical skills were more impressive than Thomson's. The German-trained Swiss chemist Jean Charles Galissard de Marignac (1817–94) was a superb laboratory chemist who by 1843 had already shown that chlorine was not the only element with an atomic weight approximating to an integral multiple of half the weight of hydrogen. He was sure that there was some truth in Prout's hypothesis. His most incisive critic was the Belgian chemist Jean Servais Stas (1813–91), also a fine analyst, who stressed that atomic weights were not generally either whole numbers or half numbers, but only near approximations to such numbers. By 1860, Marignac was publishing results to the third decimal point, and he had to admit that his numbers were not exactly what Prout's hypothesis required. Still, he argued, too many of them were too close to integers for Prout's hypothesis to be completely wrong. The numbers just had to mean something, and the idea of a prime matter was as tempting to many nineteenth-century chemists as it had been to medieval alchemists.

Before chemists could fully appreciate regularities in atomic weights and propose hypotheses about the distribution of atomic weights that would be acceptable to other chemists, they still needed to solve troubling problems. These problems had been raised by Dalton's simplicity criteria when applied to atomic weights, and by Berzelius's emphasis on reactions and the electrical nature of atoms as the guide to formulas. They remained unresolved throughout the 1850s, and even at the end of that decade chemists were still disagreeing about the atomic weights of many metals—not about decimal points, but about whether weights should be halved or doubled. In other words, they needed to adopt the solution that Gerhardt and, supporting him, Laurent had already proposed: the adoption of Avogadro's hypothesis and of "two volume" formulas.

By 1860, chemists had begun to accept the homologous formulas that Gerhardt had proposed. Determination of molecular weights using gas analysis and vapor density measurements made it easier to think in terms of the volumes that Avogadro worked with. Studies of dissociation had explained some of the apparently glaring exceptions to Gay-Lussac's law, removing another objection to Avogadro's hypothesis. Chemists were much better prepared than they had ever been for a reconsideration of Avogadro's hypothesis. An Italian chemist, Stanislao Cannizzaro (1826–1910), had arrived at this reconsideration by 1858, when he wrote a paper showing how Avogadro's hypothesis resolved a lot of problems in the determination of atomic and molecular weights and how it helped to bring together work on atomic heats and vapor densities. He also clarified the distinction between atoms and molecules. His paper, published in Italian, was at first ignored. Then he presented his argument and distributed his paper as a pamphlet at a conference in Karlsruhe, Germany. Some chemists were immediately persuaded, others read Cannizzaro's paper on the train going home and were persuaded by the time they got to their destination; and others, of course, missed the point. That conference was the turning point for the acceptance of Avogadro's hypothesis, almost a half-century after Avogadro first proposed it. Now that hypothesis could bring order to the whole of chemistry.

Mendeléev's Periodic Table

Mendeléev's periodic table was the ancestor of all subsequent periodic tables of the elements. He invented the arrangement of elements in groups (horizontal rows identified in this version of the table by Roman numerals) and periods (vertical columns, identified by Arabic numbers).

Some aspects of Mendeléev's table may at first glance confuse those who are familiar with the modern periodic table. First, here and in his early tables (1869), periods were represented as vertical columns and groups as horizontal rows, the reverse of later practice; Mendeléev subsequently changed to the later form. Second, the inert gases were missing, because they were then unknown. When they were discovered, Mendeléev placed them in group 0, at the lefthand side of the table, lacking in this illustration, whereas today we place them in group VIIIA.

We place hydrogen as the first element in the first period, along with helium. When helium was discovered, Mendeléev put it in the second period. We put the triads of iron, cobalt, and nickel; ruthenium, rhodium and palladium; and osmium, iridium, and platinum in group VIIIB, in the middle of the table. Mendeléev put them in group VIII. We also have two long groups, the lanthanides and actinides, that were a headache for Mendeléev.

There are gaps in Mendeléev's table, some that he recognized as corresponding to hitherto undiscovered elements, and others to radioactive elements, many of which are artificial and short-lived. But his table represents one of the great advances in the understanding and systematization of chemistry, and it is an essential tool in teaching chemistry.

■ D. Mendeléev, *The Principles of Chemistry*, 3rd English trans. (London, 1905), 1: xvi.

TABLE II
Periodic System and Atomic Weights of the Elements
(Giving the pages on which they are described)

	2nd Series Typical elements	4th Series	6th Series	8th Series	10th Series	12th Series
I.	Li 7 — vol. i. 574	K 39 — vol. i. 558	Rb 85 — vol. i. 576	Cs 133 — vol. i. 576	—	—
II.	Be 9 — vol. i. 618	Ca 40 — vol. i. 590	Sr 88 — vol. i. 614	Ba 137 — vol. i. 614	—	—
III.	B 11 — vol. ii. 60	Sc 44 — vol. ii. 94	Y 89 — vol. ii. 93	La 138 — vol. ii. 93	Yb 173 — vol. ii. 98	—
IV.	C 12 — vol. i. 388	Ti 48 — vol. ii. 144	Zr 91 — vol. ii. 146	Ce 140 — vol. ii. 93	? 178	Th 232 — vol. ii. 148
V.	N 14 — vol. i. 223	V 51 — vol. ii. 194	Nb 94 — vol. ii. 197	?Di 142 — vol. ii. 93	Ta 183 — vol. ii. 197	—
VI.	O 16 — vol. i. 185	Cr 52 — vol. ii. 276	Mo 96 — vol. ii. 290	—	W 184 — vol. ii. 290	U 239 — vol. ii. 297
VII.	F 19 — vol. i. 489	Mn 55 — vol. ii. 303	? 99	—	—	—
VIII.		Fe 56 — vol. ii. 317	Ru 103 — vol. ii. 369	—	Os 192 — vol. ii. 369	
		Co 59 — vol. ii. 353	Rh 103 — vol. ii. 369		Ir 193 — vol. ii. 369	
		Ni 59.5 — vol. ii. 353	Pd 106 — vol. ii. 369		Pt 196 — vol. ii. 369	

	3rd Series	5th Series	7th Series	9th Series	11th Series
I.	H 1 — vol. i. 129 — Na 23 — vol. i. 533	Cu 64 — vol. ii. 398	Ag 108 — vol. ii. 415	—	Au 197 — vol. ii. 442
II.	Mg 24 — vol. i. 590	Zn 65 — vol. ii. 39	Cd 112 — vol. ii. 47	—	Hg 200* — vol. ii. 48
III.	Al 27 — vol. ii. 70	Ga 70 — vol. ii. 90	In 114 — vol. ii. 91		Tl 204 — vol. ii. 91
IV.	Si 28 — vol. ii. 99	Ge 72 — vol. ii. 124	Sn 119 — vol. ii. 125		Pb 207 — vol. ii. 184
V.	P 31 — vol. ii. 149	As 75 — vol. ii. 179	Sb 120 — vol. ii. 186		Bi 209 — vol. ii. 189
VI.	S 32 — vol. ii. 200	Se 79 — vol. ii. 270	Te 125 — vol. ii. 270		
VII.	Cl 35.5 — vol. i. 459	Br 80 — vol. i. 494	I 127 — vol. i. 496		

Note.—Two lines under the elements indicate those which are very widely distributed in nature; one line indicates those which, although not so frequently met with, are of general use in the arts and manufactures.

Mendeléev and the Periodic Table: A Russian Revolution

By the 1860s, chemists had a set of atomic weights that were self-consistent, based on accurate analyses, and generally accepted. There were no more disagreements about whether the atomic weight of oxygen was 8 or 16. All agreed that it was 16. The kinds of regularities that had attracted Döbereiner's attention earlier in the century now began to strike other chemists. John Newlands (1837–98) in England arranged elements according to their equivalent weights and numbered them sequentially, hydrogen 1, lithium 2, and so on. He found that the eighth element after any given one "is a kind of repetition of the first, like the eighth note of an octave in music."* Thus chlorine (number 15) is the eighth element after fluorine (number 7), and both are members of the halogen group; sodium (number 9) is the eighth element after lithium (number 2), and both are alkali metals, and so on. Because of Newlands's reference to music, this numerical regularity coinciding with analogies in properties is known as *Newlands's law of octaves.*

During the 1860s, other chemists throughout Europe numbered the elements and arranged them in tables, wrote them on a line wrapped around a cylinder, plotted them on graphs, and pursued a dozen ingenious ways of displaying regularities in atomic weights or equivalent weights. Their aim was to show how such regularities brought out groupings or analogies in chemical properties. The most important and the most influential of these attempts to construct the perfect table to display the elements was by a Russian, Dmitri Mendeléev, who was born in Siberia in 1834 and died in St. Petersburg in 1907.

Mendeléev was trained as a teacher. He had published a textbook on organic chemistry in 1861. While writing it, he had been struck by one of the characteristics of Gerhardt's homologous series: within a series, physical properties and molecular weights are related, for example, the density of paraffins increases with their molecular weights. In 1868, while engaged in writing a textbook for his students, he looked for a way of classifying the elements and wondered whether something analogous to the homologous series relationship might work for elements. He became convinced that the mass of an atom and the properties of an element had to be related. By this date, around sixty elements were known, more than double the number in Lavoisier's day. Mendeléev had been to the Karlsruhe conference, and he was able to benefit from the resulting consistency in determining atomic weights. He began to write down each element on its own card, together with its atomic weight (he rejected Prout's hypothesis), its properties, and analogous elements. Then he

*Quoted in J. R. Partington, *A History of Chemistry* (London: Macmillan, 1964), 4: 887.

looked for the best arrangement of the cards, the arrangement that would most fully bring out analogies in properties and relate them to atomic weights. He concluded that the properties of the elements were in periodic dependence upon their atomic weights. By periodic he meant regular and recurring. This was the origin of the periodic table of the elements, which has evolved and grown since Mendeléev's time, but which still appears in every classroom where chemistry is taught. And of course the periodic table is based on Mendeléev's periodic law: "The properties of the elements, as well as the forms and properties of their compounds, are in periodic dependence or (expressing ourselves algebraically) form a periodic function of the atomic weights of the elements."*

Mendeléev's periodic table worked remarkably well, and he revised and improved it. Elements were ranked in increasing order of their atomic weights; the rows were arranged in columns that, read horizontally, brought out groupings of analogous elements. Alkali metals, alkaline earth metals, halogens, and other groups of elements slotted beautifully into the table. Where necessary, Mendeléev left blanks, so as to keep known elements in positions that corresponded to their chemical properties. Then, with remarkable confidence, he predicted that those blanks would later be filled by hitherto undiscovered elements, and he went on to predict the atomic weights and the chemical natures of the "missing" elements. There was a blank after zinc, in the same group as boron and aluminum. Mendeléev predicted that this blank would be filled by an element with properties similar to those of aluminum and having an atomic weight of 68 and specific gravity of 6.0. In 1875 the missing element was discovered, called gallium, with atomic weight 69.9 and specific gravity 5.96. Other elements that he predicted and that were equally close to his predictions include scandium and germanium. The periodic table not only brought out groupings and regularities among the elements, which made and makes it a valuable teaching tool, but it also encouraged chemists to look for missing elements. In many cases, they found them.

There were problems, for example, with pairs of atomic weights that were out of order for the representation of periodic properties. In Mendeléev's first periodic table, iodine (atomic weight 127) clearly belonged with the other halogens, following fluorine, chlorine, and bromine. Tellurium (atomic weight 128) belonged at the end of the sequence oxygen, sulfur, and selenium. Using atomic weight as the guide, tellurium would have come after iodine in the table, which would have meant that both elements were in the wrong groups,

*D. Mendeléev, *The Principles of Chemistry*, 4 vols. (New York: Collier, n.d.), 3: 17.

considered in the light of chemical and physical properties. So Mendeléev reversed the order of atomic weights and suggested that the weight for tellurium was wrong; it should have been less, not more, than that of iodine. He was right to be guided by the analogy of properties, and he put both elements in what we know to be the right place on the table. He was wrong in seeking to correct the atomic weight for tellurium, since it turned out to have been accurate. But without twentieth-century knowledge of atomic structure and isotopes, he had to resolve conflicting evidence, and he chose to be guided by chemical and physical analogies. In so doing, he made the decision that any good chemist would have made—the periodic table was too good to ruin.

10 The Birth of the Teaching-Research Laboratory

Until the French Revolution of 1789 and the Napoleonic Wars at the turn of the eighteenth and nineteenth centuries, chemistry in the universities had generally led a marginal existence. Most university teachers of chemistry were there to provide a service for students of medicine and pharmacy. The number of significant research chemists could be reckoned as a few dozen internationally, and, with the partial exceptions of France and Germany, it made little or no sense to talk about national chemical communities. There were distinguished professors, for example Hermann Boerhaave in the Netherlands at the beginning of the eighteenth century and Joseph Black in Edinburgh at the end of that century. For the most part, however, university chairs in chemistry were few and had little prestige. Chemistry, unlike medicine, did not constitute a profession in its own right.

There were industries that were based on the application of chemistry, but most of these depended on a traditional mixture of ingredients: entrepreneurial skill, recipes that had been found to work, and the tactile expertise of the practitioner rather than the theoretical insights of the academic chemist. Chemists were of course engaged in practical applications of their science. In the Royal Academy of Sciences in Paris, members of the Academy functioned in part as a scientific civil service and bent their energies to solving problems of water quality, street lighting, sewage disposal, and more. The French Enlightenment's great *Encyclopedia* was directly concerned with learning from the practice of artisans, and thereby both enriching theoretical understanding and improving craft and industrial practice. Joseph Black advised the masters of ironworks, Swedish chemists became expert mineralogists and consultants to the mining industry, and military chemists worked in many nations on the improvement of gunpowder. But in every one of these cases, chemistry was a tool, a servant not a master, in the view of patrons and the public if not in the view of the chemists themselves.

Chemistry lacked prestige, and chemists often worked in isolation, with little recognition from the wider community of science. Newtonian physics and astronomy were the model sciences for the eighteenth century. Many shared the great eighteenth-century philosopher Immanuel Kant's view that chemistry was incapable of becoming a science and could never be more than a kind of systematic natural history, an organized compilation of facts derived from experiment and observation. Chemistry—socially, professionally, economically, and scientifically—was a poor relation in the hierarchical family of the sciences.

The picture changed radically in the nineteenth century. By midcentury, the number of chemical chairs in Europe was in the hundreds, the status of chemistry as an autonomous discipline was secure, chemistry was increasingly prestigious in the universities, chemical societies sprang up around Europe, chemical journals multiplied as chemical research proliferated, and university-trained chemists were more and more needed in new and expanding chemical industries. Chemistry had become one of the prime engines of economic prosperity and national strength. Britain had been the leading industrial nation in the eighteenth century. By 1800, France was first in the chemical stakes, only to be overtaken by Germany after 1850. In this chapter, we shall look at the institutions of chemical teaching and research, and see how the German model came to be the dominant one in Europe. By century's end, it was also to be the dominant one in the United States.

England, or, Amateurs and Apprentices

In 1830, Charles Babbage, now mainly remembered for his invention of a proto-computer, wrote an angry book entitled *Reflections on the Decline of Science in England, and on Its Causes.* Babbage was angry because the scientific establishment, notably the Royal Society of London, had not taken enough notice of his own merits and achievements. Government did not support science in England the way it was beginning to in Germany and France. There were too few jobs for scientists, patronage was capricious and biased, and all in all, Babbage considered that science in England was in a bad way. Not everyone in England agreed with Babbage, but there was a good deal of substance to his criticisms. What he had to say about science as a whole could quite properly have been applied to the case of chemistry.

There had been isolated British chemists of distinction in the eighteenth century. We have encountered some of them, most notably Joseph Priestley and Joseph Black. Priestley had taught in a dissenting academy, and the members of the private and informal Lunar Society of Birmingham had supported his research. He had had to teach himself the techniques of research and to devise a good deal of his own apparatus. He did not have students or assistants

Experiments on Respiration at the Royal Institution of Great Britain

Humphry Davy's career began with less dignity than it later acquired. Davy came to the Royal Institution fresh from Dr. Thomas Beddoes's Pneumatic Institution in Bristol, in the west of England. Beddoes was a former student of Joseph Black in Edinburgh, a friend of James Watt and Josiah Wedgwood, an admirer of Joseph Priestley, and, like Priestley, a democrat in his politics. Having been driven out of Oxford for his political opinions, Beddoes set up shop in Bristol, and he commenced a research program to explore the therapeutic benefits of breathing different gases. He put Davy in charge of his chemical laboratory, showing extraordinary confidence in his young employee. Davy did first-rate work, especially on the effects of nitrous oxide, which we also know as laughing gas and which has been valued as an anesthetic.

In this cartoon, Davy has not yet become professor of chemistry; he is seen holding the bellows, while Dr. Garnett, a generally unsuccessful lecturer, experiments upon Sir John Hippesley, treasurer of the Royal Institution and, as an M.P., a supporter of Catholic emancipation. Since Davy's association with Beddoes connected him with radical politics and Hippesley's liberal opinions made him suspect, this cartoon can be taken as a lampoon of the political associations of pneumatic chemistry. The presence of black members of the audience also makes the connection with the antislavery movement. Clearly, pneumatic medicine and pneumatic chemistry could be seen as dangerous signs of political dissent.

■ "Scientific Researches!—New Discoveries in PNEUMATICKS!—or—an Experimental Lecture on the Powers of Air," cartoon by James Gillray, 1802, reproduced in T. H. Levere, *Chemists and Chemistry in Nature and Society 1770–1878* (Aldershot, Hampshire, and Brookfield, Vt.: Variorum, 1994), essay VIII facing p. 9.

to extend and carry on his work, and, although he was elected to the Royal Society and to foreign academies, he did not receive institutional support or recognition for his research.

Joseph Black taught chemistry as part of the medical curriculum at the University of Edinburgh. A local bottle factory made much of his laboratory glassware. Few of his students had chemistry as their principal interest. To none of them did he teach research, nor did he encourage them to publish. He was, however, in touch with James Watt and other industrialists and entrepreneurs, and his work on heat was important in the development of Watt's steam engine. One of his students, Thomas Beddoes, taught chemistry in Oxford briefly before moving to Bristol in the west of England, where he founded a medico-chemical institution. His main contribution was to provide Humphry Davy's first scientific employment, which liberated Davy from an apprenticeship to an apothecary. Davy went on to a distinguished chemical career in the privately funded Royal Institution of Great Britain. Although he did much valuable research and gave public lectures, Davy's only advanced chemical teaching was by example, in the supervision of (usually) one assistant at a time. Michael Faraday was Davy's most famous assistant and his successor at the Royal Institution. Previously a bookbinder's apprentice, he had been given a set of tickets to one of Davy's lecture courses, taken careful notes, and, by way of introducing himself, presented a bound set of those notes to Davy. A blacksmith's son, Faraday effectively became Davy's pupil and apprentice.

It will by now be obvious that chemistry in Britain around 1800 was not concentrated in the universities and that even where it was taught, the lecturers and professors did not teach research. Science was not seen as a profession, and this certainly applied to chemistry. This was not a matter of accident. It was entirely deliberate. In a court case later in the nineteenth century, chemists were denied fees as expert witnesses because they did not agree about the evidence. This, according to the judge, proved that chemistry was not a science and that chemists as witnesses were simply members of the public, not expert professionals. Honors for scientists were few and far between, and for chemists, exceedingly rare. Humphry Davy became a baronet, but he was alone among English chemists of his generation in receiving this title.

The professions in England in 1800 were in the church, the army, the law, and medicine, and at this date were exclusively for men. Where chemistry was taught at the universities, it was either as a part of the medical curriculum or to satisfy general interest. When academic standards at Oxford were tightened in the early nineteenth century, students had to pay more attention to the subjects needed for their degree, and their attendance at chemistry lectures all but

Michael Faraday Lecturing at the Royal Institution

While Humphry Davy's lectures had made chemistry fashionable, Michael Faraday, his successor at the Royal Institution, did more than any other English chemist of his generation to broaden what we now call the public understanding of science. His lectures were so popular that the average attendance was more than the safe limit for the lecture theater of the Royal Institution. From 1835 until 1862, when he stopped lecturing, Faraday gave lecture courses for students and around seventy-five formal Friday evening discourses. He also took the leading role in establishing the Christmas Lectures for children, giving nineteen sets of these lectures himself. Best known among these is his "Chemical History of a Candle," first given in 1848 and published in 1861. Since then the lectures have been repeatedly, indeed almost constantly, in print. Faraday's earliest researches, carried out under Davy's guidance, were in different branches of chemistry, but when the newfangled word *scientist* was invented in 1834, Faraday insisted on retaining for himself the older term, *natural philosopher.* Certainly his electrochemical researches, which form a most impressive sequence from the 1830s, belong to chemistry and to physics alike.

In this illustration, Faraday is lecturing to a most respectable audience. Prince Albert sits in the center of the front row, flanked by two of the royal princes. Albert, German born, was very much a patron of the sciences. He was a key figure in the Great Exhibition of 1851, in the establishment of science museums and colleges, and in the encouragement in England of German chemists.

■ Faraday lecturing before the Prince Consort in 1855. From a lithograph by Alexander Blaikley, reprinted in G. Porter and J. Friday, eds., *Advice to Lecturers: An Anthology Taken from the Writings of Michael Faraday and Lawrence Bragg* (London: Mansell for the Royal Institution, 1974).

vanished. The ancient universities of Oxford and Cambridge did not offer laboratory instruction and had no interest in turning out qualified practical chemists. Where they taught chemistry, they did so as part of a liberal education, an education that might include mathematics and even some natural philosophy but was not primarily geared toward the sciences. A liberal education was not meant to be a practical or a directly useful affair but was directed instead at the inculcation of habits of thought and of a common culture appropriate for gentlemen. When London University was founded in the 1820s, it was partly to redress the imbalance and to give the sciences, including chemistry, a more significant place in education. Even then, chemistry was not a prominent part of the curriculum; that had to wait until the foundation of the Royal College of Chemistry in London in 1845. The model for this new college was German, not British, and its first professor was also German.

It is significant that the extensive and deliberate application of chemistry to industry in Britain began in the industrial midlands. There, religious dissenters including Quakers and Unitarians, excluded from the old universities, had set up their own colleges where chemistry was taught. In their communities, as in the Scottish Enlightenment, making money from the application of chemistry in a technological and entrepreneurial context was a respectable and worthy activity.

France after Napoleon: Paris or Bust

Paris was the center of power in France, but in the case of the sciences before the end of the eighteenth century, Paris was not the be all and end all. Montpelier, for example, was a rival to Paris in medicine, and its professors had genuine prestige. Following the French Revolution, and most strikingly under Napoleon's rule, centralization became the order of the day, and a job outside Paris scarcely counted. In the first half of the nineteenth century, there were only seven chemistry chairs in all the provincial French universities. At mid-century, Paris had more than half the chemistry chairs in France, and they were the ones that carried prestige. It took government action in the 1850s to begin to restore provincial universities to health.

But if the provinces suffered, chemistry in Paris became extremely healthy. Chemists in Paris became leading figures in national life. Gay-Lussac was a professor at the Ecole Polytechnique, founded in the aftermath of the French Revolution. He was also a peer of France, a director of the mint, as well as an active politician. Other chemists became government ministers, members of the Legion of Honor, directors of industrial enterprises, and even friends and advisers to Napoleon. In the middle of the nineteenth century, the most influential chemist in France was Jean-Baptiste Dumas (1800–1884), who had two professorships in Paris, was minister of agriculture and commerce from 1849 to 1851, then minister of education, as well as a figure of power and authority in the Paris Academy of Sciences.

The stature of individual chemists was not, however, identical with the health of chemical education and research in France. The Ecole Polytechnique became under Napoleon a scientific training school for French army officers, and it had some of the most brilliant chemists of the day teaching there. They did not, however, teach research, nor did they have research teams working under them. Their job was simply teaching. They, like the professors of chemistry in the universities, had heavy teaching loads, and they did not receive funds for research laboratories. Many of the leading chemists in France in the generation after Lavoisier did their most impressive research in the context of the private Society of Arcueil. Arcueil, then a village and now a suburb of Paris, was where the physicist Laplace, one-time co-worker of Lavoisier, and Claude-Louis Berthollet (1748–1822), the chemist whose support was decisive for acceptance of Lavoisier's new theory, established their remarkable society. They had a laboratory and a journal, and they exercised extraordinary patronage amounting to control of chemistry and physics in France. France became the leading European nation in chemistry in the years around 1800. Among the reasons for this dominance were the success of Lavoisier's chemical revolution, including the adoption of French chemical nomenclature; Napoleon's enthusiasm for chemists; and the research and patronage of the members of the Society of Arcueil. That dominance was not to last.

Two reasons for France's relative decline in chemistry were the numbing effect of centralization in Paris and the lack of regular university funding for the teaching of research, which mostly took place outside the universities. Dumas's research school in Paris is the most striking exception to this generalization, and he had worked with Justus Liebig, the German inventor of a new kind of institution, the teaching laboratory, where the techniques of research could be learned. Germany provided the model to which French chemists referred in their attempts to find funds for research. These attempts finally bore fruit in the 1860s and 1870s, the years before and after the Franco-Prussian War, in which the French saw German scientific superiority, especially in chemistry, as a major cause of German victory and French defeat.

Germany, Liebig, and the Virtues of Government Interference

Justus Liebig's career was made possible by the structure of the German university system, which achieved its modern form thanks to the Prussian reorganization of universities in the first decades of the nineteenth century. In 1807, during the Napoleonic Wars, the philosopher Johann Gottlieb Fichte (1762–1814), dissatisfied with the state of education, drew up plans for a new kind of university at Berlin. Two years later, Karl Wilhelm von Humboldt (1767–1835), at that time Prussian minister of education, was one of the founders of the new University of Berlin. The institution was committed to the disciplined systematic study of all realms of knowledge, including chemistry and the other natural sciences. Its most innovative feature was its commitment to a combination of teaching and research. Professors were expected to carry out research and to contribute to the continuing growth of knowledge by educating the next generation of research students; this was not only an innovation but also a contrast with the separation of teaching from research that characterized the old established English universities. Philosophy, the inquiry into the nature and limits of knowledge, provided a context for all fields of knowledge. Thus, at Berlin, chemistry and the other sciences belonged in the faculty of philosophy. That was the rationale for the German invention of the modern form of the Ph.D. degree, in chemistry and all other disciplines of knowledge. Academic freedom, where students could choose their programs of study and professors could pursue research topics of their own choice, was an explicit commitment of the new university.

In the years that followed, Berlin was the model for the foundation of other new German universities and for the radical reconstruction of old ones. Germany was not yet a unified country but rather a set of German-speaking states and provinces, each of which came to see its own university as a source of prestige. So French-style centralization was neither desirable nor possible. Besides the universities, the Germans also pioneered the technical high school, a university-level institution for such practical scientific and technological studies as engineering, including, toward the end of the nineteenth century, chemical engineering. Chemical engineering was a response to the limits of industrial chemical production and led directly to improved yields and greater efficiency. The technical high schools, like the universities, offered research degrees, and their graduates had a social status comparable with that of university graduates. That status matched the prestige of graduates of the Ecole Polytechnique in Paris and contrasted strongly with the lower status of the purveyors of practical knowledge in early nineteenth-century England.

The German model gradually spread internationally. American universities imported the model, reforming old institutions and, in the late nineteenth century, founding new universities with a dual mandate in teaching and research. The first to import the German model of graduate research was the Johns Hopkins University, which opened in 1876. English universities, where teaching and research had formerly been distinct activities, followed suit by awarding doctoral degrees to research students much later; the Ph.D. degree was first given in England in 1919.

Prussian reorganization of higher education provided the background for Liebig's career. So too did the existence of training schools for apothecaries and pharmacists. The most impressive of such schools in Germany had been founded at Göttingen in 1795. Liebig, with an early competence in chemistry and with the model of such training schools before him, planned further studies in chemistry in Paris, after which he would come back to Germany and open his own institute with a teaching laboratory for training pharmacists and chemists. In Paris, Liebig attracted the attention of Gay-Lussac, who after a while invited Liebig to carry out researches under his direction. This was a flattering invitation and a wonderful opportunity, and Liebig grabbed at it, extending his stay in Paris by a year, learning the techniques of research in Gay-Lussac's own laboratory and acquiring the kind of dedication that shows itself in very long work hours and generates impressive results. In Paris, Liebig discovered his potential as a research chemist. He also wrote a joint paper with Gay-Lussac which did his career no harm at all.

When he returned to Germany, he was ambitious and wanted an academic post at one of the more impressive universities. What he got instead was an appointment as an extraordinary professor (more or less our associate professor) at what he considered a lesser university, Giessen, and he obtained that post only because the naturalist, traveler, and diplomat Alexander von Humboldt (1769–1859) put a lot of pressure on the local patron and potentate, the grand duke of Hesse. This was done without consulting the professors already at Giessen, who were predictably annoyed.

Liebig did however find some helpful colleagues, and with them, he founded precisely the kind of training school for pharmacists and chemists that he had in mind before his studies in Paris. But the standards that he had learned in Paris required abilities in chemical analysis that went beyond what such a training school could impart, and so Liebig began to reshape his school to produce organic analysts of a high caliber. His own standards, learned from Gay-Lussac, impelled him to create a new kind of institution, in which the techniques of analysis served as prelude to the pursuit of research. Liebig's was the first university laboratory that prepared students for research and for membership in a research school. Organic analysis, the prime focus of Liebig's

At Work in Liebig's Laboratory at Giessen in 1842

Liebig pioneered the teaching-research laboratory and, as the top hats and elegant jackets worn by the students in this illustration suggests, made the practice of chemistry a dignified and respectable profession. He was one of the main players in the establishment of organic chemistry as a subdiscipline, and he went beyond academia to bring chemistry into general repute and the widest awareness. His work on agricultural chemistry and on animal chemistry, published in popular form and soon translated from German into English, showed the importance of chemistry in the production of food and in the understanding of bodily health. He also made significant contributions to public health. His laboratory was an outstandingly successful supplier of academic and industrial chemists in Germany and England. The model of his laboratory spread to the United States, where, later in the nineteenth century, the Johns Hopkins University offered the first North American Ph.D. program in chemistry.

■ From a lithograph by W. Trautschold and H. von Ringen, reproduced in W. H. Brock, *Justus von Liebig: The Chemical Gatekeeper* (Cambridge: Cambridge University Press, 1997), 55.

school, was a good choice of field, since inorganic analysis was already a fine and exact art, led by Berzelius. Organic analysis was less developed, and so offered more opportunity.

When it came to small organic molecules, the techniques already in use were adequate, since a small percentage error was unlikely to translate into an error in the number of atoms. That meant that existing methods of analysis, including those that Liebig had learned in Paris, would probably generate the right formula for a compound. But as chemists began to investigate larger and larger molecules, with many carbon atoms and more hydrogen atoms, a small percentage error could mean that the wrong number of atoms would be deduced, and so the wrong formula would be obtained. This was clearly unsatisfactory.

The best analytical chemists, including Berzelius and Gay-Lussac, could already produce results of impressive accuracy. In the years following his appointment at Giessen, Liebig worked at developing methods of organic analysis that were rapid and reliable. His most dramatically successful innovation in apparatus was his potash apparatus for determining the weight of carbon dioxide, the gaseous combustion product of carbon, which enabled him to work out the carbon content of an organic substance. It was not restricted, as earlier techniques had been, to the analysis of a few tenths of a gram of an organic sample but rather could be used for much larger quantities. Working with larger samples made for greater precision in measurement.

The apparatus, although small, was simple and powerful. It consisted of five glass bulbs connected by glass tubes. The bulbs were filled with an aqueous solution of potash (potassium hydroxide, KOH), which absorbed carbon dioxide, forming potassium carbonate. The arrangement of five bulbs ensured that all the carbon dioxide was absorbed and that no splashes escaped. The apparatus and its contents were weighed before and after the experiment, and the recorded increase in weight led to an accurate estimation of carbon from the substance being analyzed.

Although one major advantage of the apparatus was that it could be used for very precise analyses of large samples, the apparatus was also sensitive enough for good results to be obtained from small samples. This was important when many newly discovered compounds, whether extracted from natural products or synthesized in the laboratory, were often available only in small amounts. With this apparatus, accurate measurements were possible even for relatively unskilled chemists, analyses were more rapid than before, and Liebig found his own researches, and those of his students, made faster progress. Given the proliferation of organic compounds in Liebig's laboratory and elsewhere, accurate analyses were increasingly urgent, and Liebig's potash apparatus met that need. He borrowed other apparatus from different chemists, including Berzelius and Gay-Lussac. For the estimation of nitrogen he relied on techniques and apparatus invented by Dumas.

Armed with these methods, Liebig and his students between them had a dominant role in the emerging field of chemistry. From the mid-1830s, he attracted more and better students internationally, and he offered a well-structured course in laboratory practice in which practical training and research blended

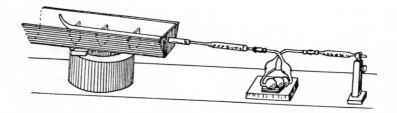

Liebig's Apparatus for Determining Carbon Dioxide

The essential device, Liebig's potash apparatus, is the central part of this experimental arrangement. The size of the device is about 10 cm in height or width. When an organic compound is burned in oxygen, carbon dioxide is one of the products. Gas coming through the tube on the left can be strongly heated in the first part of the apparatus, then passed over a drying agent, which absorbs any water vapor. The dry gas then goes through the potash apparatus, where carbon dioxide is absorbed by the potash solution. The difference between the weights of the potash apparatus before and after the experiment gives the amount of carbon dioxide, from which one can easily calculate the amount of carbon produced in the combustion of the organic compound being analyzed. This device, which now provides the logo for the American Chemical Society, was cheap and easy to produce, needing no great skills in glass blowing.

■   From J. von Liebig, *Anleitung zur Analyse organischer Körper* (Braunschweig, 1837), plate 1, reproduced in F. Szabadvary, *History of Analytical Chemistry* (Yverdon: Gordon & Breach, 1992), 293.

into each other. Low fees, thorough training, and Liebig's remarkable charisma as a teacher and director combined to make his laboratory an international magnet for the best students. From 1824 until Liebig left Giessen in 1852, more than seven hundred students matriculated in chemistry and pharmacy in his laboratory. By 1839, Liebig could teach thirty students at a time in the laboratory. Four years later, he was able to add another laboratory in which fifteen more students worked under his senior assistant. Between 1830 and 1850, thirteen of Liebig's German students became full professors of chemistry and twelve achieved the same rank in technical high schools. Five became full professors of technical and pharmaceutical chemistry. Among his foreign students in the same period, twenty-nine became full professors of chemistry at a university; four, at technical high schools. British chemistry came to be dominated by the Liebig school at Giessen. Two of his foreign students became full professors of pharmacy. Liebig's students filled more chairs of chemistry than did the students of any other professor or school in these years, at home and abroad. They took many of the top jobs in British academic chemistry.

It is worth noting that ninety-three of Liebig's students went into industrial chemistry, and another fifty-three became consulting chemists. Seventeen went into technical government departments. Liebig's laboratory was clearly not only a very efficient breeding ground for practitioners of the rapidly expanding world of academic research chemistry and a very successful training school for pharmacists. It was also a prime supply of chemists for industry and commerce. Indeed, the number of students armed with training in analysis and in chemical research who found employment in industry shows that academic research in organic chemistry was of clearly perceived value to manufacturing industry and related commerce.

To the traditional industries of mining and metallurgy—the production of mineral acids (sulfuric acid production achieved particular importance), the production of alkalis, and bleaching—were added both minor and major new industries. These included the manufacture of phosphates as plant food, as recommended by Liebig, and the petroleum and coal-tar-related industries, rich in opportunities and challenges to organic chemists and serving as foundations for the manufacture of a host of organic dyestuffs derived from coal tar. August Wilhelm von Hofmann (1818–92), Liebig's most significant pupil in consolidating and extending the model of the teaching-research laboratory, was also, through his work on coal tar and its derivatives, a key figure in the establishment of the German industry of aniline dyes. He is the perfect example of the linkage between university research and industrial production. Organic chemistry was well poised to contribute to economic growth in the mid-nineteenth century and beyond, and Liebig and his students were major participants in the science that underlay this growth.

Applied chemistry was based in part on "pure" chemistry, and Liebig, although well aware of the importance of applying his science, was a strong advocate of curiosity-driven or pure research. Chemical science became recognized as a key to material wealth and progress, important to nations as well as to business. That is why governments were prepared to fund Liebig's teaching laboratory.

Liebig had done what all successful entrepreneurs need to do. He found a niche in the market that was important but not yet adequately filled—organic chemistry, and especially organic analysis. He acquired and invented the equipment (laboratory space and apparatus, including his five-bulb potash apparatus) necessary for the effective exploitation of this niche. He found the necessary financial backing from the state and the university, showing an impressive return on investment. He advertised skillfully, that is, he published extensively, created his own journal, and encouraged his students to publish. And he built

Liebig in the Marketplace and in the Kitchen

When Liebig took chemistry to the people, he did it not only through his writings in popular science but also through advertising products that were nutritionally sound. This advertisement for the Liebig Company's beef extract includes the advice, "Note this Signature in BLUE on every Jar," immediately above what must have been the most widely recognized signature in the history of chemistry. The ad appears in Mrs. Beeton's cookery book.

■   Advertisement in Mrs. Isabella Beeton, *The Book of Household Management . . . ,* new edition (London, 1898).

FOR

GOOD COOKERY

USE

LIEBIG
COMPANY'S EXTRACT

A pure concentrated essence of the finest beef, its use in the preparation of gravies, soups, sauces, meat pies, and all savoury dishes imparts to them the essential features of good cookery — appetising, flavour, nourishment, and digestibility. Nothing can take its place.

Note this Signature in BLUE on every Jar.

AVOID INFERIOR SUBSTITUTES

---

an international network, based on personal knowledge. International research students and visiting researchers, as well as graduates who took the reputation of the Liebig school wherever they went, extended this network. These activities ensured a constant supply of ever-stronger students, the essential workforce for his industry.

In the 1820s and 1830s, Liebig systematically introduced his students to the techniques of organic analysis and then supervised their independent research in which they used the techniques he had taught them. This was the very model of Ph.D. education, now almost universally adopted. But by 1840, Liebig began to direct his students so that several or even many of them worked on different aspects of a single coordinated research project—Liebig's own research project. In the 1840s, the main work given to Liebig's students was to provide analytical information for Liebig's work in agricultural chemistry, fermentation, and animal physiology. Students had to analyze the ash produced by burning plant matter, in order to support Liebig's views on the nature and role of minerals in plant physiology. Liebig's interest in the chemistry of digestion and nutrition led his students to undertake analyses of proteins as well as of animal excrement and urine. This was a kind of chemical account book or balance sheet of input and output, food and excreta, as well as of gases breathed in and out by animals and humans, in a line of research that went back to Lavoisier. Analytical chemistry was the key, the tool rather than the goal of all this work, and Liebig made sure that his students were skillful in analysis. Liebig became particularly interested in the role of muscles, and he marketed a meat extract with a label showing a vigorous bull over the signature "Justus von Liebig" (the "von" was the result of Liebig's ennoblement in 1845).

After 1850, Liebig spent less time as director of research and more in popularizing the results of the researches that he and his students had pioneered. Agricultural chemistry, which he virtually reinvented, and animal chemistry were the subjects of two of his best-selling books, and English translations sold extremely well in England and the United States. He also entered politics and spent a good deal of time supporting his former students and their careers. The last phase of his career took him increasingly away from the laboratory. The earlier phases, based in the university chemical laboratory, had created the essence of the teaching-research laboratory, which, together with the teaching-research seminar, formed the main institutions that led to German leadership in nineteenth-century science. They also provided models for the development of science over the next hundred and fifty years. Liebig's contributions were not unique—Dumas also created a powerful research school—but they produced the first institutional teaching-research laboratory and have had enormous influence and consequences.

11 Atoms in Space

Chemists in the 1840s and 1850s had spent a good deal of time considering the problem of the arrangement of atoms in molecules. Laurent insisted that molecules with formulas belonging to the same homologous series had analogous properties *because* their atoms were arranged in the same way. He believed that molecular properties owed more to the position of atoms than to their nature. He argued that when substitution took place, converting one substance into another with similar properties, the similarity in properties was caused by similarity in the arrangement of atoms. Thus, when chlorine replaced hydrogen so as to convert acetic acid into trichloracetic acid, chlorine atoms literally took the place, the spatial location in the molecules, of the substituted hydrogen atoms. There was, however, as Laurent's co-worker and friend Gerhardt insisted on pointing out, a problem. There was no way of knowing what the actual arrangement of atoms was. Indeed, when pushed by Gerhardt, Laurent went so far as to admit that it might prove forever impossible to know precisely how atoms were arranged in molecules. In chemistry, as in every science, there were things that were known, things not yet known that were knowable in principle, and things unknowable in principle.

Radical theory, which had its first dramatic success in a paper on the benzoyl radical by Liebig and Wöhler in 1832, suggested other ways in which atoms were grouped, and electrochemical dualism seemed, in spite of Laurent's criticisms, to offer insights into arrangement and chemical bonds. Nonetheless, in the 1840s, it seemed entirely possible that the arrangement of atoms in space might belong to the category of the unknowable in principle. But what is unknown and considered unknowable in one generation of scientists may shift to the category of the knowable, or even of the known, in another generation. That is what happened in the case of the position of atoms in molecules. This change required a complete about-face by chemists, and so it is not surprising that it was accompanied by vigorous and sometimes angry debate. There was a lengthy prelude, followed by two dramatic announcements by the Dutch

chemist Jacobus Henricus van't Hoff (1852–1911) and the French chemist
Joseph Achille Le Bel (1847–1930).

From Radicals and Types to Valence

Gerhardt's theory of types (see Chapter 9) used four molecules—hydrogen
(H_2 or HH), hydrogen chloride (HCl), water (H_2O or HOH), and ammonia
(NH_3)—as the models or types of all organic compounds. In hydrogen, two
like atoms are combined; in hydrogen chloride, two unlike atoms; in water,
two like atoms (hydrogen) are combined to a third dissimilar atom (oxygen);
and in ammonia, three like atoms (hydrogen) are combined with a fourth dissimilar atom (nitrogen). Gerhardt said nothing about how they were combined. He preferred to confine himself to what he knew for certain and not to
waste time speculating about what he could not know. Other chemists soon
came up with different theories of types, and not all of them were as opposed
as Gerhardt was to speculation. One of them wrote in 1852: "Formulas . . . may
be used as an actual image of what we rationally suppose to be the arrangement of constituent atoms in a compound, as an orrery is an image of what
we conclude to be the arrangement of our planetary system."*

Four years later, the German chemist Friedrich August Kekulé (1829–96),
ambitious, successful, and dominating, proposed a development of type theory in which the main issue was explicitly the number of atoms of an element
or radical that combine with another atom or radical. This combining power
was one for hydrogen, two for oxygen, and three for nitrogen; these set the pattern for all other elements. Kekulé followed Gerhardt in proposing three principal or fundamental types, hydrogen (HH), water (OH_2), and ammonia
(NH_3). Replacement of one of these atoms by another of similar combining
power resulted in molecules like HCl, SH_2 (hydrogen sulphide), and so on.
Multiple types could also arise by the combination of two or more of the fundamental types. This is complicated, but no more so than building a complex
structure out of a small number of different kinds of Lego blocks, some constructed so as to be capable of combining with only one other block, others
with two or three blocks. Elaborate buildings are likewise made out of just a
few kinds of simple bricks. Such thinking made it possible for chemists to consider a complex molecule containing many atoms as put together using simple

*Alexander J. Williamson, *Journal of the Chemical Society* 4 (1852): 351, quoted in C. A. Russell, *The History of Valency* (Leicester: Leicester University Press, 1971), 51. An orrery is a mechanical device showing the movements of the planets around the sun. It was invented ca. 1700
and named after Charles Boyle, the earl of Orrery.

architectural rules. Of course there were far more than three elements—the number of known elements by the late 1850s had more than doubled the number in Lavoisier's list of 1789. But, according to Kekulé in the 1850s, the combining power of each and every atom was one, two, or three. Each element belonged to one of three categories, characterized by combining power.

These developments in thinking about the way molecules were built from atoms fell far short of saying just what the spatial arrangement of those atoms might be. But they helped chemists to think about the links between atoms. Almost all the developments that we have considered so far in this chapter either originated in organic chemistry or found their original application in organic chemistry. Laurent, had he lived so long, would have felt thoroughly vindicated in his controversy with Berzelius about the right way to do chemistry. Laurent had told Berzelius, and had insisted all along, that chemists should explore the whole of their science using the methods and models developed in organic chemistry.

This insistence reveals an irony in type theory. The theory grew out of organic chemistry, but its fundamental types were simple *inorganic* molecules. In this respect, it can be seen as a step toward the unification of chemistry that Laurent wanted, but it seems odd to find a key theory in organic chemistry that makes no mention of the one element essential to organic compounds. That element is carbon. In 1871 Henry Roscoe wrote in his textbook of chemistry that "Organic Chemistry is defined as the chemistry of the carbon compounds."[*] The *Oxford English Dictionary* defines organic chemistry as "the chemistry of the hydrocarbons and their derivatives." Carbon is the one atom that is common to all organic substances, and type theory would not realize its full potential until carbon was brought into that theory. This happened twice in 1858, in two independent papers. The paper that proved the most influential was by Kekulé and was entitled "On the Constitution and Metamorphoses of Chemical Compounds, and on the Chemical Nature of Carbon."[†]

Kekulé paved the way for this paper a year earlier, when he added a fifth type to Gerhardt's four. The new type involved the combination of four atoms or radicals with carbon. Because the simplest molecule of this type is methane or marsh gas, CH_4, it was known as the marsh gas type. Kekulé also insisted that his types were not merely characterized by similarity of formula and prop-

---

[*]Henry Roscoe, *Lessons in Elementary Chemistry* (London, 1871), 289.

[†]The other was by Archibald Scott Couper. Kekulé claimed priority, under doubtful circumstances, and Couper promptly suffered a complete mental breakdown and made no further contributions to chemistry. Quotations in translation from Kekulé in this section are from Russell, *The History of Valency,* 61ff.

erties, as Gerhardt's were, but were related by the chemical transformations that they could undergo. One molecule of the marsh gas type could be "produced from, or transformed into" another of the same type. And, as Kekulé implied in a footnote to another paper written in 1857, the composition of molecules of this type could be explained by the nature of carbon. "Carbon," he wrote, "is . . . tetratomic: i.e. one atom of carbon . . . is equivalent to four atoms of hydrogen." In 1858, he elaborated this insight. "The amount of carbon which the chemist has known as the least possible, the atom, always combines with four atoms of a monatomic, or two atoms of a diatomic element; . . . generally, the sum of the chemical units of the elements which are bound to one atom of carbon is equal to four." Today, and indeed since the 1860s, we use the word and the concept *valence* to describe the combining power of different atoms. The valence of an element is its capacity to combine with a certain number of hydrogen atoms or of atoms with the same combining power as hydrogen. Carbon can combine with four hydrogen atoms, so it has a valence of four; it is *tetravalent*.

In the 1850s and 1860s chemists wrestled with ways of representing this idea in the formulas that they wrote. The problems were obvious. How could chemists represent the connections, links, or, as they became known in the 1860s, *bonds* between different atoms when it was difficult for them to find agreement about the arrangement of atoms in a molecule? How *were* atoms linked? Most chemists were reluctant to commit themselves to representations of what, a few years previously, would have been dismissed as the wildest speculation.

We have already encountered the irony that the theory of types was based on inorganic molecules but applied to organic molecules by chemists who argued that organic chemistry provided the analogies for uniting all chemistry. But it was not just the theory of types that gradually lost its hard edges and its simple analogies. The radical theory underwent a similar metamorphosis. We have already seen how Liebig and Wöhler in 1832 had achieved the first major triumph of radical theory, which held out promise that inorganic chemistry, and specifically the electrochemical properties of atoms and radicals, might one day account for all reactions. Over the next two decades, the radical theory underwent changes, giving increasing recognition to the place of atoms and radicals within molecules. It is therefore not surprising to find some chemists, emerging from the radical tradition, who combined the formulas of type theory with electrochemical concepts.

One of the leading chemists in this line was the English chemist Edward Frankland (1825–99), who had studied in Germany with Robert Bunsen and

Liebig before returning to England to become a professor of chemistry. Frankland argued that methyl, ethyl, and homologous organic radicals possessed the same character as hydrogen, although they were less electronegative. He used type formulas for organic and inorganic molecules. He also observed that there was a clear symmetry between the formulas of many stable compounds with well-satisfied combining powers; for example, in the case of NH_3 (ammonia), PH_3 (phosphine), AsH_3 (arsine), PCl_3 (phosphorus trichloride), and so on. To account for the symmetry, he suggested that "no matter what the character of the atoms may be, the combining power of the attracting element, if I may be allowed the term, is always satisfied by the same number of these atoms."[*] The same element tended to have the same combining power with other different elements. That suggested a particular number of bonds for a given element, and it invited representations of such bonds.

Touching and overlapping circles, touching sausage-shapes and circles, letters within circles connected by straight lines, brackets, dashes, and straight-line connections, and even ball-and-stick models were developed in the 1860s.[†] Ball-and-stick models were especially uncomfortable to contemplate, because they represented the arrangement of atoms in space, which went beyond what chemists were confident about. Some, indeed, viewed such models as fantastic and outrageous.

In contrast to what some chemists saw as the false promise of ball-and-stick models, formulas using straight lines to link atoms indicated merely the *order* in which atoms were joined. Thus methane had a carbon atom attached by single bonds to four separate hydrogen atoms. It could be represented on the page by the graphic formula

$$\begin{array}{c} H \\ | \\ H-C-H \\ | \\ H \end{array}$$

Carbon had a valence of four and could be linked to other atoms, including other carbon atoms, by more than one bond. Thus in ethane, C_2H_4, each carbon atom was joined by two bonds, or a double bond, to the other carbon atom, and by a single bond to each of two atoms of hydrogen:

[*]E. Frankland, "On a New Series of Organic Bodies Containing Metals," *Philosophical Transactions of the Royal Society of London* 142 (1852): 417–44, at 440.

[†]J. Loschmidt used touching and overlapping circles. Kekulé used sausage-shape formulas; then, after intermediate stages, letters joined by straight lines. The unfortunate Archibald Scott Couper and after him Alexander Crum Brown developed formulas using straight lines as bonds. Edward Frankland used a different system of lines to illustrate valence bonds.

$$\begin{array}{ccc} H & & H \\ \diagdown & & \diagup \\ & C = C & \\ \diagup & & \diagdown \\ H & & H \end{array}$$

In acetylene, C_2H_2, there was a triple bond between the carbon atoms, both of which used their fourth valence bond on a hydrogen atom:

$$H-C\equiv C-H$$

The formulas were all planar, that is, two-dimensional, although it was clear to many chemists by the end of the 1860s that it was very unlikely that all molecules really were planar. By 1870, Kekulé was using graphic formulas like these, as were many other chemists. Kekulé was not the only inventor of such formulas, although his energy in claiming priority for himself has given him the lion's share of the credit.

The combination of bond theory with graphic formulas and the tetravalence of carbon produced what is known as the *structural theory*. The resulting chemical formulas are commonly called *structural formulas*. One of the early successes of structural theory was resolution of the formula of benzene, C_6H_6, a substance that had lately become the focus of interest because of its importance in the newly important petroleum, coal tar, and dyestuff industries. Kekulé, here as elsewhere, was at the forefront in research and in claiming priority for his findings. He first tried sausage formulas and found that a regular distribution of hydrogen atoms resulted in two single bonds left over. The obvious thing to do was to join them together to form a six-sided ring, which he did in 1865. In the following year, he added alternating single and double bonds, giving us the modern symbol for benzene:

He subsequently claimed that the idea had come to him in a fireside dream, with six snakes joining head-to-tail in a ring.

Chemists argued about priorities, proposed different kinds of structural formulas, and were energetic in their rivalry, but they clearly formed part of a community with common goals and, usually, shared understanding of the rules. Then, in 1874, two papers were published that shocked the more conservative members of the chemical profession. These papers dared to specify the position of atoms in space.

Soaring on the Wings of Pegasus

The offending papers were published almost simultaneously in 1874. Although they were completely independent of one another and argued in very different ways, they arrived at the same conclusions. Van't Hoff had studied in the Netherlands, then worked for a while under Kekulé in Germany. Then he worked in Charles-Adolphe Wurtz's laboratory in Paris, where he met Le Bel. Le Bel had studied at the Ecole Polytechnique, the great French scientific and technical school that trained technical officers for the army.

Van't Hoff's paper was entitled "On the Relations which Exist between the Atomic Formulas of Organic Compounds and the Rotatory Power of Their Solutions," which must have seemed fanciful if not absurd to many of his contemporaries because it implied a relation between chemical formulas and what had been considered a purely physical property. Light rays may be considered as being made up of waves. When a ray is modified so that it displays different properties on different sides, it is said to be polarized. The sides of a ray can take different directions; polarizing sunglasses transmit only that portion of light with rays polarized in a particular direction. Some substances are optically active, which means that they, or their solutions, have the power of rotating the plane of polarized light. Before Van't Hoff's work, nobody had perceived any chemical significance in the phenomenon of optical activity.

Van't Hoff began by considering the phenomenon of isomerism, in which the same atoms combined differently in different substances. In 1828 Wöhler had shown that urea, an organic compound extracted from urine, had the same composition as the inorganic compound ammonium cyanate, that is, it contained the same number of the same kinds of atoms (carbon, oxygen, hydrogen, and nitrogen). The two substances had different properties, and chemists concluded that the difference arose because the atoms were differently arranged. But how were they arranged? By the 1870s, structural formulas were in general use, and Van't Hoff observed that these formulas were incapable of accounting for certain cases of isomerism in organic chemistry.

Take methane, CH_4, and progressively substitute different radicals, R_1, R_2, R_3, and R_4, for the hydrogen atoms. That gives us molecules $CHHHR_1$, $CHHR_1R_2$, $CHR_1R_2R_3$, and $CR_1R_2R_3R_4$. How many isomers of each substance will there be? If we take the structural formulas developed by Kekulé and others, then for $CHHHR_1$ we have only one possibility:

(1)

$$R_1$$
$$|$$
$$H-C-H$$
$$|$$
$$H$$

Other arrangements of these atoms, for example,

$$H$$
$$|$$
$$H-C-R_1$$
$$|$$
$$H$$

are identical with the first arrangement, as we can see by simply rotating the images of a molecule, known as graphical formulas, until they can be placed on top of one another, with R_1 on top of R_1, and so on. If two formulas can be superimposed, so that they correspond in every way to each other, then it follows that they describe identical molecules.

(2)

$$R_1$$ $$R_1$$
$$|$$ $$|$$
$$H-C-H$$ and $$H-C-R_1$$
$$|$$ $$|$$
$$R_1$$ $$H$$

Rotating *these* formulas does not enable us to superimpose one exactly on the other, so the structural formulas suggest that their atoms are differently arranged and should form two different isomers. The same is true of molecules with two like and two unlike atoms or radicals linked to the central carbon atom.

(3)

$$H$$ $$H$$ $$R_2$$
$$|$$ $$|$$ $$|$$
$$R_1-C-R_2$$ $$R_1-C-R_3$$ $$R_1-C-H$$
$$|$$ $$|$$ $$|$$
$$R_3$$ $$R_2$$ $$R_3$$

These cannot be superimposed, and therefore should, according to the formulas, constitute three different isomers.

The trouble was that no one had found isomers corresponding to those indicated by the structural formulas in (2) above, and there were not three isomers corresponding to (3) above. Theory did not match the facts. Van't Hoff observed that "the theory is brought into accord with the facts if we consider the affinities [bonds] of the carbon atom directed towards the corners of a tetrahedron of which the carbon atom itself occupies the center."* A tetrahedron is a kind of pyramid, a geometric solid with four triangular faces and four corners. If there is a central carbon atom connected by its four valence bonds to four atoms or groups, at least two of which are the same (e.g., $CR_1R_2R_2R_3$, or $CR_1R_2R_3R_3$), then it turns out that all tetrahedral arrangements of these atoms or groups correspond to molecules that can be superimposed on one another, and so the isomers predicted by the two-dimensional graphic formula in (2) are in fact not to be found. If there is a central carbon atom connected by its four valence bonds to four different atoms or groups, then there are two forms of the molecule that, no matter how one rotates them, cannot be superimposed on one another. Again, the two-dimensional graphic formula in (3) predicts three isomers, but the tetrahedral formula predicts just two, and experiment shows that there are only two such isomers.

Theories, once established, are valuable for making predictions. But Van't Hoff did not begin by inventing the theory of the tetrahedral carbon atom and then go on to predict the existence of isomers. Rather, he came up with the theory to explain the nonexistence of some isomers and the existence of others as laboratory research revealed them. The two molecules $CR_1R_2R_3R_4$ in figures VII and VIII (*opposite*) are the mirror image of each other, like left- and right-handed gloves. As Van't Hoff wrote:

> Imagining specifically the line R_1R_3, with one's head at R_1 and looking towards the line R_2R_4, R_2 can lie either on the viewer's right . . . or left . . . ; in other words: *in cases where the four affinities [bonds] of the carbon atom are saturated by four mutually different univalent [single] groups, two and not more than two different tetrahedra can be formed, which are each other's mirror images, but which cannot ever be imagined as covering each other, that is, we are faced with two isomeric structural formulas in space.*†

Carbon atoms linked or bonded to four different groups in this fashion are known as *asymmetric* carbon atoms. They form isomers that cannot be divided

---

*Van't Hoff's 1874 pamphlet, reprinted in Peter J. Ramberg and Geert J. Somsen, "The Young J. H. van't Hoff: The Background to the Publication of His 1874 Pamphlet on the Tetrahedral Carbon Atom, Together with a New English Translation," *Annals of Science* 58 (2001): 51–74.

†Van't Hoff's 1874 pamphlet, in Ramberg and Somsen, "The Young J. H. van't Hoff" (emphasis in original).

Van't Hoff's formulae, showing mirror-image symmetry (figs. VII, VIII), double bonding (figs. IX, X), and triple bonding (fig. XI) between carbon atoms.

■ From J. H. van't Hoff, *Voorstel tot uitbreeding der tegenwoordig in de scheikunde gebruikte structuurformules in de ruimte: benevens een daarmee samenhangende opmerking omtrent het verband tusschen optisch actief vermogen en chemische constitutie van organische verbindingen* (Utrecht, 1874).

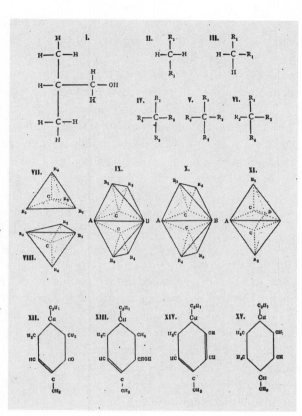

into two matching or symmetrical halves. Such molecules are called *enantiomers,** from the Greek root for "opposite"; their shapes or structures are related to each other as an object is related to its image in a mirror. They are mirror images of each other. They are also called *stereoisomers,* because they differ in the spatial arrangement (not in the order of connection) of their constituent atoms or radicals. Now it is very often hard, or even impossible, for chemists to find any strictly chemical difference between enantiomers. Living systems, in contrast, often manage to distinguish between them, or rather to produce only the left-handed or only the right-handed form of particular substances, such as some of the alkaloids that make certain mushrooms highly poisonous.

There is however one difference between enantiomers that is easy to identify in the laboratory, and that is their effect on polarized light. Some chemical substances in solution rotate the plane of polarized light, as polarized sun-

*They are sometimes called *enantiomorphs,* from the Greek for "opposite forms or shapes."

glasses (not in solution) would do if they were themselves rotated. Van't Hoff determined that "all of the compounds of carbon which in solution rotate the plane of polarized light possess an asymmetric carbon atom,"* and he found in this asymmetry the explanation of their isomerism. Left- and right-handed enantiomers rotate the plane of polarized light in opposite directions.

So far we have considered only carbon compounds in which single bonds link the central carbon atom to four different atoms or radicals. Van't Hoff's tetrahedrons also work for double bonds between carbon atoms, when two tetrahedrons have an edge in common (see figs. IX and X, *p. 145*), and for triple bonds, when two tetrahedrons have a triangular face in common (see fig. XI, *p. 145*). Molecules containing such bonds can still be optically active if they possess the necessary overall asymmetry in the form of left- or right-handedness.

Van't Hoff's seminal paper was originally published as a pamphlet in Dutch, then translated into French in a French-language Dutch journal. Van't Hoff at the time was working at the veterinary school in Utrecht in the Netherlands. In a Europe dominated by German chemistry, but in which French and English chemists also had significant authority, employment in a "cow college" in the Netherlands and publication in Dutch did not give Van't Hoff's ideas the instant visibility and recognition that they deserved. Instead, they attracted ridicule, most notably from Hermann Kolbe (1818–84), a leading but conservative organic chemist who was then a professor at Leipzig University. Kolbe complained that modern chemistry was suffering from the most unfounded speculations by frauds whose theories were as out of place as "freshly painted harlots" in the respectable salons of scientific society.

> Whoever thinks this worry exaggerated should read, if he is capable of it, the recent phantasmagorically frivolous puffery [of Van't Hoff. He] finds, it seems, no taste for exact chemical research. He has considered it more convenient to mount Pegasus (apparently on loan from the Veterinary School) and to proclaim in his ["Chemistry in Space"] how, during his bold flight to the top of the chemical Parnassus, the atoms appeared to him to be arranged in cosmic space. The prosaic chemical world has no time for these hallucinations.†

Van't Hoff was sufficiently taken with the idea of soaring on Pegasus that he reprinted the attack in the second edition of his paper. But it was hard for conservative chemists in the 1870s to see the point of a theory that drew on no direct *chemical* evidence. They regarded optical activity as simply a physical

---

*Van't Hoff's 1874 pamphlet, in Ramberg and Somsen, "The Young J. H. van't Hoff."

†Quoted in translation in Alan J. Rocke, *The Quiet Revolution: Hermann Kolbe and the Science of Organic Chemistry* (Berkeley and Los Angeles: University of California Press, 1993), 329.

property of some substances. Logic and physics combined were not enough to convince all chemists that Van't Hoff's argument enabled them to know where atoms really were in a molecule. As we shall see in the next chapter, Van't Hoff went on to make a hugely successful career from the combination of chemistry and physics, and he became the first Nobel laureate in chemistry.

Le Bel did not publish in Dutch, but rather in the *Bulletin of the Chemical Society of France,* which gave him an instant advantage over Van't Hoff. He also came at the problem of stereoisomerism differently. Van't Hoff began by thinking about the number of isomers predicted by a formula, and then showed how formulas needed to be extended into three dimensions to make theory and laboratory experience agree with one another. Le Bel, in contrast, began with optical activity. Louis Pasteur and others had already shown the connection between asymmetry and optical rotatory power. Certain crystals (e.g., quartz crystals) are asymmetric, that is, they have no plane of symmetry. Pasteur painstakingly picked out the left- and right-handed forms of such crystals, separating them from one another, and showed that they possessed equal and opposite powers of optical rotation. As Le Bel summarized this and later work: "If the asymmetry exists only in the crystalline molecule [i.e., in the solid crystal], the crystal alone will be active; if, on the contrary, it belongs to the chemical molecule the solution will show rotatory power, and often the crystal also."[*]

Then, using considerations that were strictly geometric, Le Bel arrived at the general rule that any molecule with four different atoms or radicals attached to a central carbon atom would be optically active unless it had a plane of symmetry. He used this rule to discuss various kinds of optically active organic molecules. The lack of a plane of symmetry in molecules $CR_1R_2R_3R_4$ is only possible if the carbon atom has four bonds directed to the corners of a tetrahedron. So, effectively, Le Bel arrived at the same conclusion as Van't Hoff. But he drew no diagrams representing three-dimensional models, and his argument was more abstract and general than Van't Hoff's. Van't Hoff, and not Le Bel, has received most of the credit for the creation of organic formulas.

These formulas, once accepted, led to a huge increase in the intelligibility and predictability of organic reactions. Chemists could now work out, in their notebooks or on the blackboard, the structures of new substances and their probable reactions. They could also devise on paper ways of making such substances. Then, by laboratory experiment and demonstration, they very likely would be able to confirm their predictions. With knowledge of the structure

[*]Le Bel, "On the Relations which Exist between the Atomic Formulas of Organic Compounds and the Rotatory Powers of their Solutions" (1874), reprinted in O. T. Benfey, *Classics in the Theory of Chemical Combination* (New York: Dover, 1963), 161.

of benzene and of its derivatives (the field known as *aromatic* chemistry) and
with an understanding of the tetrahedral carbon atom, organic chemistry, once
perceived as a dark and threatening forest, became a brightly lit arena for the
confident extension of chemical knowledge. The last quarter of the nineteenth
century saw a dramatic expansion of organic chemical industry, particularly
the dyestuffs industry. The theoretical innovations of Kekulé (especially the
benzene ring and structural formulas) and of Van't Hoff and Le Bel (tetrahedral carbon) made major contributions to a chemical industrial revolution.

Also of great importance was the rapid development of chemical synthesis, which accelerated throughout the second half of the nineteenth century.
Sometimes, using the techniques of Liebig, Dumas, Frankland, Wurtz, Kolbe,
and Marcelin Berthelot (1827–1907), it seemed that if one could extend a homologous series on paper, one could also extend it in the laboratory. As valence
theory developed, along with an understanding of the stability of the benzene
ring, the invention of molecules on paper was matched only by the rapidity of
discoveries and new syntheses in academic and industrial laboratories. Berthelot claimed that he had practically invented organic synthesis, which was not
at all true; but he had identified one of the most successful fields of chemical
endeavor, the foundation of the dyestuffs industry in the nineteenth century
and of the pharmaceutical industry in the twentieth century.

Alfred Werner and Coordination Chemistry

Organic chemistry was riding high. Theory and practice reinforced each other,
profitable applications proliferated, and so did jobs for organic chemists, in industry and in universities and colleges. Inorganic chemistry, which under
Berzelius had been the dominant branch, was beginning to look like a poor relation. Two nineteenth-century contributions to inorganic chemistry started
to turn this situation around, although the real rebirth of inorganic chemistry
was in twentieth-century Australia. The first of the pivotal contributions in the
nineteenth century was Mendeléev's periodic table, bringing order to the classification of elements and making possible the prediction of new elements.
The second was Alfred Werner's (1866–1919) invention of *coordination chemistry*, an extended application of valence theory and stereochemical thinking
to the chemistry of metals. Werner, born in France, did his doctoral work in
Switzerland and then worked in Paris. After his return to Switzerland, he
taught at the University of Zurich until his death. His work on coordination
chemistry earned him the Nobel Prize for chemistry in 1913.

In his Ph.D. thesis, a joint project with Arthur Hantzsch, Werner showed
that trivalent nitrogen (i.e., nitrogen with three valence bonds) could form

stereoisomers. Van't Hoff's reasoning about carbon, that different spatial arrangements lead to different isomers, could be extended to nitrogen. These three formulas have the same numbers of atoms, joined to the same atoms and groups, but arranged differently in space:

$$C_6H_5-C-----------C-C_6H_5$$
$$\quad\quad \| \quad\quad\quad\quad\quad \|$$
$$\quad\quad N-OH \quad HO-N$$

$$C_6H_5-C-----------C-C_6H_5$$
$$\quad\quad \| \quad\quad\quad\quad\quad \|$$
$$\quad\quad HO-N \quad\quad HO-N$$

$$C_6H_5-C-----------C-C_6H_5$$
$$\quad\quad \| \quad\quad\quad\quad\quad \|$$
$$\quad\quad HO-N \quad\quad N-OH$$

This, while new and important, was not revolutionary, at least where valence was concerned. Werner was working with existing theories, showing how they could be extended.

His next step, presented in 1893 in his paper "Contribution to the Constitution of Inorganic Compounds," was revolutionary, with insights of startling originality and productivity. Significantly, it was published in a new German journal for inorganic chemistry. Werner concentrated in this paper on compounds of cobalt and platinum, but his theoretical arguments had much wider applicability.

So far, we have considered valence bonds only as fixed in number for a given element and, as Van't Hoff had shown for carbon, as having particular directions. Werner, like some but by no means all other chemists, rejected both of these notions, productive though they had proved. As far as direction was concerned, he regarded valence as an attractive force acting uniformly from the center of the atom over its entire surface. He was convinced that valence bonds could move; they did not have fixed directions.

Next came a rejection of the common interpretation of *molecular compounds*. If nitrogen had a valence of three, as in ammonia, NH_3, and if this never varied, then how could chemists explain the formation of what we call ammonium chloride, NH_4Cl, which seemed to imply a combining power of five units, that is, a valence of five? Similarly, how did phosphorus, which was generally viewed as having a valence of three, form PCl_5? One answer proposed was that the molecules of ammonia and hydrogen chloride had an attraction for one another, forming the molecular compound NH_3,HCl. Phosphorus trichloride could similarly form the molecular compound PCl_3, Cl_2. Werner rejected this notion, adopted the idea of variable valence, and argued that the atoms, radicals, or groups added when molecular compounds involving metals were formed were joined directly to the metal.

Thus, for example, the compound whose elemental composition corresponded to the formula $PtN_2H_6Cl_2$ was not the molecular compound $Pt(NH_4Cl)_2, Cl_2$, but rather had the ammonia (NH_3) molecules and the chlorine atoms bound directly to the platinum atom. There were two isomeric forms of this substance, and Werner proposed that the formula should be planar, producing the two isomers:

$$
\begin{array}{ccc}
& NH_3 & \\
& | & \\
Cl - & Pt & - Cl \\
& | & \\
& NH_3 &
\end{array}
\qquad\qquad
\begin{array}{ccc}
& NH_3 & \\
& | & \\
NH_3 - & Pt & - Cl \\
& | & \\
& Cl &
\end{array}
$$

A tetrahedral platinum atom was not possible in these molecules, since the isomers were not optically active. The direct linking to the metal of the non-metallic groups added in the formation of these supposed molecular compounds was known as *coordination* and the resulting molecules were *coordination compounds*. Molecules that we consider coordination compounds today include hemoglobin and chlorophyll, which are vital to animal and plant life.

The planar square molecular structure turns out to be common among coordination compounds. The other common structure is octahedral, where the metal is at the center of an eight-sided geometric solid with six vertices, formed by joining two square pyramids at their base. For example, a platinum atom can be linked to four chlorine atoms and two molecules of ammonia in this fashion. If the four chlorine atoms are at the corners of the square, and the ammonia molecules at the apexes, then they form one optical isomer. If one of the ammonia molecules and three of the chlorine atoms form the square, and one chlorine atom and one ammonia molecule occupy the apexes, then we have another optical isomer. The two forms are enantiomers of one another:

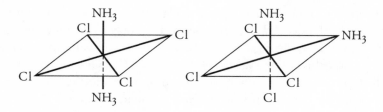

It was not immediately easy for chemists to get into the habit of visualizing chemical compounds in three dimensions and to see how the same groups could be arranged in different ways. But, once they had made this jump, in inorganic and organic chemistry, they could see how molecules might interact, how groups with an affinity or attraction for one another might be brought together, and what reactions might take place. They could also see how parts of a molecule might be more or less accessible, because of the directions of bonds and the extent to which geometric factors increased or decreased the accessibility of a group to would-be reactants. Organic chemistry after Van't Hoff became an arena where theory could predict and precede experiment. Inorganic chemistry after Werner made the same step.

Once chemists had arrived at the beginnings of an understanding of the electronic nature of valence (see Chapter 13), the way was open to a detailed prediction of the mechanism of chemical reactions. In organic chemistry especially, an understanding of structure and of the nature of the chemical bond combined to turn the theoretical branch of the science into a kind of predictive gymnastics, where structure and mechanism were interpreted interdependently. One of the early twentieth century's leading organic chemists, Christopher Ingold (1893–1970), made major contributions to "the pulling together of organic chemistry" in such a fashion. His first such contribution was in a paper entitled "Principles of an Electronic Theory of Organic Reactions"; the second was in his magisterial and influential text, *Structure and Mechanism in Organic Chemistry* (Ithaca, N.Y.: Cornell University Press, 1953; 2nd ed., 1969). Anyone who enjoys chemical puzzles in three dimensions will find this a challenging and exciting book.

12 Physical Chemistry

A Discipline Comes of Age

Chemists since the time of Robert Boyle have had widely varying attitudes toward the place of chemistry in relation to physics. But that is a misleading way of putting things, since chemistry as a discipline emerged convincingly during the eighteenth century, whereas the discipline of physics came later. What we now call physics was known in the seventeenth century as natural philosophy, of which mechanical philosophy was a major strand. Chemists in the seventeenth and also in the eighteenth centuries had repeatedly to reconsider the relations between natural philosophy and chemistry. As the discipline of physics emerged in the nineteenth century, chemists had to reckon with this new aspect of the old natural philosophy.

As we have learned, the philosopher Immanuel Kant held that chemistry could never be a real science; it would forever be limited to functioning like natural history, a science of classification but not of explanation. Kant had a prejudice in favor of Newton's mathematical physics and astronomy and against what he considered to be lesser sciences. Classification has certainly been an important part of the business or discipline of chemistry, but it has never been the whole story. Although some chemists displayed hostility to mechanical philosophy, others made a practice of weaving parts of mechanical philosophy and the wider natural philosophy into their chemical inquiries. Boyle, for one, examined the compressibility of air, what he called the *spring* of the air, and saw that investigation, like his laboratory chemistry or chymistry, as part of mechanical and natural philosophy. In the late eighteenth century, heat was part of the subject matter of chemistry and calorimetry, the measurement of quantities of the supposed matter of heat, was part of chemical practice. In the following century, heat became part of the science of physics, while chemists studied thermal phenomena that accompanied chemical change. In the 1800s, the voltaic pile attracted the attention of many chemists, and chemists began to explore *electrolysis,* the chemical action of electricity. In the 1830s Michael Faraday discovered some of the laws of electrolysis. The study of electricity itself, like heat, became part of physics, while chemists explored electrochemistry. In the 1840s, Laurent and Gerhardt used the physical properties of organic compounds to guide their classification, and in the 1860s Mendeléev used the physical properties of elements to reinforce his periodic classification.

The existence of these different practices was not sufficient to create a discipline or subdiscipline of physical chemistry, but it showed the way. One definition of physical chemistry is that it is the application of the techniques and theories of physics to the study of chemical reactions, and the study of the interrelations of chemical and physical properties. That would mean that Faraday was a physical chemist when engaged in electrolytic researches. Other chemists devised other essentially physical instruments and applied them to chemical subjects. Robert Bunsen (1811–99) is best known today for the gas burner that bears his name, the Bunsen burner, a standard laboratory instrument. He also devised improved electrical batteries that enabled him to isolate new metals and to add to the list of elements. Bunsen and the physicist Gustav Kirchhoff (1824–87) invented a spectroscope to examine the colors of flames (see Chapter 13). They used it in chemical analysis, to detect minute quantities of elements. With it they discovered the metal cesium by the characteristic two blue lines in its spectrum and rubidium by its two red lines. We have seen how Van't Hoff and Le Bel used optical activity, the rotation of the plane of polarized light (detected by using a polarimeter) to identify optical or stereoisomers. Clearly there was a connection between physical and chemical properties.

Hermann Kopp, one of Liebig's students, moved from a chair at Giessen to one at Heidelberg, where he became that university's first professor of physical chemistry. He aimed to show that all physical properties were determined by chemical composition. He was one of the first to occupy a chair in physical chemistry. Kopp's goal was an ambitious and all-embracing one. For the most part, however, the activities and investigations listed above may fit into the dictionary definition of physical chemistry but do not amount to a coherent discipline. It seems reasonable to agree with the claim that physical chemistry is a *dis*unified practice; its practitioners were brought together by their fragmented successes.

Three Musketeers

Kopp had his chair of physical chemistry at Heidelberg from 1863 on. Another scientist, Gustav Wiedemann, more a physicist than a chemist, had a laboratory at Leipzig starting in 1871. He sought to construct a foundation for "general chemistry," which was another name for what became known as physical

chemistry. (In Heidelberg, "physical chemistry" was already in use.) But neither Kopp nor Wiedemann succeeded in creating the discipline of physical
chemistry, with its journals, research schools, and a community of practitioners. Three chemists, from three different countries, are generally credited with
bringing physical chemistry into being as a discipline. The chemists were Van't
Hoff from the Netherlands, Wilhelm Ostwald (1853–1932) from Latvia, and
Svante Arrhenius (1859–1927) from Sweden. All three won Nobel Prizes in the
first decade of the twentieth century for their work in physical chemistry, beginning with Van't Hoff in 1901, the first year that a prize was awarded in chemistry. Their triple success was a clear recognition that physical chemistry had
reached maturity. Nobel Prizes are awarded only when the significance of a particular piece or program of research has become evident. By 1909, when Ostwald received his Nobel Prize, the world of chemistry had fully recognized
physical chemistry as an established subdiscipline.

Ostwald was a student at the University of Dorpat (Tartu) in Estonia, one
of the Baltic states. He then went to teach at the Polytechnic Institute in Riga
in his native Latvia. There in the 1880s he continued the work that he had
started in Dorpat on the theory of chemical affinity. Affinity, one of the main
concepts in eighteenth-century chemistry, had during the nineteenth century
lost its centrality. Ostwald was determined to modernize it and bring it back
into the mainstream of chemistry. In the mid-1880s, he became more ambitious. Especially after his appointment to the chair of physical chemistry at
Leipzig in Germany in 1887, he decided to build new foundations for the whole
science of chemistry. He believed that nineteenth-century chemists had come
close to deserving Kant's criticism and that they spent too much time as tax-
onomists, identifying and classifying individual substances just as naturalists
identified and classified species of living organisms. Organic chemistry, in particular, had in Ostwald's opinion concentrated excessively on classifying compounds. Chemists were so busy learning about the substances that took part
in chemical reactions, the events during which chemical substances were
changed and entered into new combinations, that they virtually ignored the
reactions themselves. That was Ostwald's growing conviction, and he wanted
to refocus chemists on those reactions.

He wanted answers to questions that had received little attention. How did
changes in temperature and pressure affect the way in which reactions took
place? Reactions could occur in principle in two different directions, from reactants to products and from those products back to the original reactants.
Such *reversible* reactions were common, achieving equilibrium when the reactions in each direction had stabilized and balanced one another. What was the

nature of that equilibrium, and how was it determined? How did chemical affinity and the mass of reacting substances influence or determine the nature of a reaction and the equilibrium point? How could one predict the yield of a reaction, a significant economic issue for the industrialist? What determined the rate at which reactions took place?

Descriptive answers would no longer be enough. Ostwald wanted to determine the general laws of chemical change, using physical techniques as appropriate, and present these laws in the language of mathematics and analysis. In the study of continuously changing magnitudes, such as the concentration of reactants in a reversible reaction, the appropriate mathematical language is calculus.* The study of the transfers of heat and energy in dynamic systems is called *thermodynamics,* and it was to be applied to chemical reactions in the newly developing field of *chemical thermodynamics.* The study of reaction velocities is called *chemical kinetics.* The nature of solutions, and the behavior of substances capable of carrying an electric current in solution, also turned out to be important for these questions. Together, these topics and questions form only part of what we now regard as the discipline of physical chemistry, but they were crucial and central components in the growth of that discipline.

Ostwald, Arrhenius, and Van't Hoff would have welcomed any and every extension of physical chemistry that came their way and every subsequent enlargement of their discipline. After all, physical or general chemistry, in the minds of its founders, would be the basis for the whole science of chemistry. It would then become general chemistry. With such an outlook, chemical spectroscopy and quantum chemistry (see Chapter 13) were simply two more areas to be welcomed and incorporated into physical chemistry. Physical chemists had no trouble extending their field into new areas, and no difficulty with what some may have seen as a blurring of the boundaries between physics and chemistry. As one distinguished modern physical chemist observed, "Nobel Prizes in chemistry have sometimes gone to people (such as Ernest Rutherford and Gerhard Herzberg) who thought that they were physicists, and vice versa. The distinction between the sciences is, after all, no more than a matter of administrative convenience."†

If the first decade of the twentieth century, with its three Nobel awards to

---

*Calculus is generally seen by those who are unfamiliar with it as a forbidding and difficult part of mathematics. A healthy antidote to that phobia is the book by Sylvanus P. Thompson, *Calculus Made Easy: Being a Very-Simplest Introduction to Those Beautiful Methods of Reckoning which Are Generally Called by the Terrifying Names of the Differential Calculus and the Integral Calculus,* rev., ann., and exp. by Martin Gardner (New York: St. Martin's Press, 1998), which appeared in its original edition in 1910 and has as its motto What One Fool Can Do, Another Can.

†Keith J. Laidler, *The World of Physical Chemistry* (Oxford: Oxford University Press, 1993), 5.

Van't Hoff, Arrhenius, and Ostwald, marked the complete recognition of physical chemistry, 1887 marked the beginnings of its recognizable status as a discipline. In that year, Ostwald and Van't Hoff founded the German-language *Journal for Physical Chemistry* (*Zeitschrift für physikalische Chemie*), published in Leipzig, where Ostwald was a professor. The initial papers were of major significance and included one of Van't Hoff's essays on chemical thermodynamics, the subject for which he won his Nobel Prize, as well as pioneering work by Arrhenius on electrolytic dissociation. The editorial board was unprecedentedly and impressively international, including distinguished chemists from Britain and around Europe, and even Mendeléev from Russia. Although this was the first journal in the field, it was by no means the last. Other European journals appeared in the 1890s. In 1896, Wilder Bancroft, who, after studying at Harvard University, worked in the laboratories of Van't Hoff and Ostwald and obtained his Ph.D. at Leipzig in 1892, founded the *American Journal of Physical Chemistry*. The discipline was international and firmly established during the 1890s and 1900s.

It is time to see how Ostwald, Van't Hoff, and Arrhenius developed their researches and came together in publicly affirming the establishment of their discipline. In doing so, we shall not attempt to cover the whole of physical chemistry but shall concentrate on those questions and areas that brought the three principal performers together.

Van't Hoff and Thermodynamics: You Can't Win

Van't Hoff, after writing a relatively trivial Ph.D. thesis in organic chemistry, devoted the rest of his research career to pursuing his conviction that much of chemistry could be reduced to physics. We have already seen the first fruit of this conviction in his groundbreaking paper on the tetrahedral carbon atom. Neither that paper nor his thesis initially did him much good. He suffered almost two years of unemployment before obtaining a frustrating position at the Veterinary School in Utrecht. Hermann Kopp, criticizing Van't Hoff's paper on tetrahedral carbon, poked fun at Van't Hoff's lowly position. In 1878 Van't Hoff became a professor at the University of Amsterdam, and there began his researches in thermodynamics and chemical kinetics in order to explain chemical equilibrium and chemical affinity. In 1896 he moved to Berlin, where he was able to devote himself uninterruptedly to research through a professorship at the Berlin Academy of Sciences, soon coupled with a chair at the University of Berlin that did not require him to give lecture courses. In Berlin he worked on the equilibrium of salts in marine salt deposits (the study of chemical equilibrium is central to chemical thermodynamics). Before we can see how Van't Hoff applied thermodynamics to chemical systems, we need to have at least a qualitative grasp of some of the key concepts and laws of thermodynamics as they were worked out by physicists.

Thermodynamics generally considers the relations between heat, work, temperature, and energy in systems in or near a state of equilibrium. Such systems, whether isolated from their environment or exchanging matter and energy with that environment, tend toward a state of equilibrium. That state is stable and can be described in terms of such properties as temperature and pressure. If the system undergoes change, for example if it changes its volume, then the properties of the state of equilibrium will also change. Thermodynamics describes such changes (e.g., when a gas undergoes a sudden change of volume and pressure), and thermodynamics also predicts the properties of a system in equilibrium. Thermodynamic descriptions are couched in mathematical language. Mathematical physicists developed the science of thermodynamics without concerning themselves with chemical change and the ways in which different kinds of chemical substances behave; physicists are interested in determining general laws. Chemists, however, took up thermodynamics and extended and applied it to the study of chemical reactions and chemical equilibrium.

The beginnings of thermodynamics came in the early nineteenth century, inspired by the need to determine the efficiency of steam engines and the optimal conditions for their use. Three laws were developed. The *first law of thermodynamics* is simply the law of the conservation of energy. Energy can be converted from one form into another, for example, when chemical batteries produce electric currents, or when the mechanical motion of magnets and wires in alternators produces electricity, or when fuel is burned to produce work and heat. Every automobile on the road has an alternator and a battery and obtains its power from the combustion of fuel, so driving an automobile involves chemistry and different forms of energy. But no matter how complex the interconversion of different forms of energy, the total amount of energy in the universe remains constant. When it comes to energy, you cannot get something for nothing, or, using the metaphor of a game of chance, you may not lose but you cannot win.

The earliest thermodynamic work on the efficiency of heat engines showed that there was a predictable theoretical limit to the efficiency of such engines. No matter how well built and designed they might be, some heat was always lost in the conversion of heat into mechanical work. This was the fundamental insight behind the *second law of thermodynamics*. Again, in terms of our imaginary game of chance, not only can you not win, you cannot break even.

There were many different ways in which the second law was expressed, and it is not always obvious that they are mathematically equivalent. William Thomson, Lord Kelvin (1824–1907), stated the law this way in 1851: "It is impossible, by means of inanimate material agency, to derive mechanical effect from any portion of matter by cooling it below the temperature of the coldest of the surrounding objects." In 1865 Rudolf Clausius (1822–88) expressed the first and second laws of thermodynamics as follows: "The energy of the universe is constant; the *entropy* tends towards a maximum."* *Entropy* was a term invented by Clausius, and it became absolutely central to the understanding and expression of thermodynamics.

Entropy is a mathematically defined quantity that can be interpreted as a measure of the order or disorder of a system—if the entropy increases, then so does disorder; if it decreases, then disorder also decreases. It can also be seen as a way of gauging the usefulness of energy. Ordered energy is useful; disordered energy is not. The energy in a tank of gasoline, or in any other fuel, is relatively ordered; when the gasoline has been burned and used up, the energy it contained has been spread far and wide, has become more disordered, and is no longer accessible or useful. It is because entropy tends to a maximum, that is, tends to increase, that there is a limit to the efficiency of heat engines. If two bars of gold are heated to different temperatures and are then brought into contact, heat flows from the hotter bar to the cooler one until together they reach the same temperature, in a state of equilibrium. Once they are in equilibrium, no further net transfer of heat will take place. Beforehand, when one bar was hot and the other cold, there was more order than when both came to the same temperature. Equilibrium occurs when entropy is at its maximum. Because of the operation of the second law of thermodynamics, the entropy of a closed, nonliving system will never decrease spontaneously, so the process that leads to equilibrium is irreversible.

Temperature relates heat to entropy. As a body is heated and raised to a higher temperature, its particles move more violently, as in boiling water, and there is more disorder, more entropy. Conversely, as a body is cooled, temperature and entropy both decrease. There is a theoretical minimum temperature, −273°C, known as absolute zero, where the parts (e.g., atoms) of a body are at perfect rest and perfectly ordered, so that entropy is at its theoretical minimum. The *third law of thermodynamics* states that as a body is cooled and approaches absolute zero, the further extraction of heat (energy) becomes harder and harder, so that however close one gets, it is impossible to reach absolute

*Kelvin and Clausius quoted in Laidler, *The World of Physical Chemistry,* 100, 105.

zero. A simpler way of stating the third law is that entropy changes become zero at absolute zero. In terms of our game of chance, the third law means that you cannot get out of the game.

Chemical reactions involve transfers of energy and the absorption or emission of heat. They may be reversible, for example when all reactants and products remain in solution from the moment when the reactants are mixed to the point where equilibrium is reached. As an example of a *reversible* reaction, consider the reaction between hydrogen and iodine to form hydrogen iodide, $H_2 + I_2 = 2HI$. The reaction can take place in both directions, from hydrogen and iodine to hydrogen iodide, and from hydrogen iodide to hydrogen and iodine; at equilibrium, chemical change takes place at equal rates in the two opposite directions. For an example of an *irreversible* reaction, consider the mixture of solutions of silver nitrate with sodium chloride, to form silver chloride and sodium nitrate: $AgNO_3 + NaCl = AgCl + NaNO_3$. Instead of a dynamic equilibrium between four *ions*,* $Na^+ + (NO_3)^- + Ag^+ + Cl$, sodium nitrate and silver chloride are formed. The silver chloride is insoluble in water, so it forms a precipitate that falls out of solution. The reaction continues until no more silver chloride can be formed, and then the reaction stops. Sodium nitrate, in the form of sodium and nitrate ions, remains in solution. In reversible reactions, the rate at which equilibrium is reached, the balance of reactants and products, and the nature of the products are all a function of temperature, pressure, concentration, and the chemical nature of the substances involved. Van't Hoff and later chemists used chemical thermodynamics to explore the way in which chemical equilibrium is reached, the factors affecting the direction of a chemical reaction, and the conditions of equilibrium.

Van't Hoff was not the first to do important work in chemical thermodynamics. That distinction belongs to the American Josiah Willard Gibbs (1839–1903), who was appointed professor of mathematical physics at Yale University in 1871. In the 1870s he published three papers on the importance of entropy. He proposed a function (now known as *Gibbs energy*, formerly Gibbs free energy) that has to decrease if a reaction is to take place spontaneously. This Gibbs energy is determined by the internal energy of a system and also by pressure, temperature, and volume. Gibbs also worked out the conditions necessary for reaching chemical equilibrium. These were all extremely important contributions to chemical thermodynamics, but they were put forward in abstract mathematical form, without reference to specific chemical reactions or systems, and were understood by very few of his contemporaries. Intelligi-

---

Ion was the name given by Michael Faraday to the components of a compound which pass to either the positive or negative pole (electrode) in electrolysis. Salts are composed of ions.

bility and presentation are crucial for the successful reception of even the most important ideas and discoveries; through neglecting these aspects, Gibbs won little reputation in his lifetime. There is a story, which may be true, that the president of Yale said to a visiting British physicist that the university was looking for a professor of theoretical physics. When told that Yale already had Gibbs, the president replied that he had not heard of him. Only in the 1930s did Gibbs's work gain major recognition and influence.

Van't Hoff certainly did not know of Gibbs's work until after his own major researches; nor did any other chemist working on thermodynamics in the 1870s and 1880s. Van't Hoff began his thermodynamic work while teaching at the University of Amsterdam, in spite of a heavy teaching load. He tackled thermodynamics as a chemist interested in making sense of chemical reactions. He wanted to discover simple equations that he could relate and apply directly to his laboratory work on specific substances and their reactions. He devised equations to describe the variation of the concentration of different reactants with time. He thought about the work done by chemical affinity in a reaction. He found that equilibrium depended upon temperature, and he showed how that temperature dependence worked. Indeed, he worked out equations to describe the equilibrium of any chemical system, although there were practical difficulties in some specific applications. His work drew together the more fragmented contributions of several predecessors and added to them. He published his researches on the thermodynamic treatment of dynamic chemical equilibrium in many papers in several journals, including the one that he and Ostwald founded in 1887. He effectively summarized his work in *Studies in Chemical Dynamics* (1st ed. in French, 1884; 2nd ed. in German, 1896).

Solutions Are Problems: Arrhenius's Solution

We have mentioned salts in solution (i.e., dissolved in water), reversible reactions and equilibrium in solution, and ions in solution. Most chemical reactions occur in solution. It was apparent to Van't Hoff at an early stage that to understand the dynamics and thermodynamics of chemical reactions, he needed to understand the nature of solutions in general. And it was equally clear when he began his work that very little was known about the nature of solutions. Solutions always involve specific chemicals, the solvent (often water) and the dissolved substance or solute (often a salt, e.g., sodium chloride). But although they are always chemical systems, they can also be considered as physical systems, to which the principles of thermodynamics can be applied.

One aspect of reactions in solution which Van't Hoff treated, and which we have already mentioned, concerned dynamic equilibrium and the way this depended on temperature. Another area that he explored was that of osmosis and osmotic pressure. If a porous membrane separates pure water from an aqueous solution of a salt, then water will flow through that membrane so as to dilute the solution. This process is called osmosis. The pressure needed to prevent the entry of pure water into the solution is called osmotic pressure. Osmosis is a physical phenomenon, and it obeys laws analogous to the laws relating the temperature, pressure, and volume of a gas. As a physical phenomenon, osmosis could be tackled using the second law of thermodynamics, and Van't Hoff did so. At the same time, he interpreted osmotic pressure in terms of the chemical affinity of a dissolved substance for water. In Van't Hoff's initial work on solutions, his mathematical predictions did not match up with observations of chemical systems in solution. The analogy between gases and solutions was a powerful one, but unfortunately the predictions derived from that analogy worked only when the solution was infinitely dilute. That could have been an indication that Van't Hoff, and Ostwald, who was also working on the theory of solutions, were completely on the wrong track, or it could have indicated that their theory of electrolyte solutions was incomplete.

Time and again in the history of science, scientists confronted with a mismatch between prediction and observation have had the choice of jettisoning their theories, or modifying those theories and reinterpreting the data, or seeking new and better data. The middle course, reinterpreting the data and modifying rather than rejecting the theory, has often proved to be the right one. So it was for Van't Hoff, Ostwald, and the thermodynamics of salt solutions. What was lacking in their earliest work on solutions was an understanding of the dissociation of salts in solution.

This gap was filled by the young Swedish chemist Svante Arrhenius, who provided the necessary modification and extension of Van't Hoff's as well as Ostwald's theory of solutions. In 1884 Arrhenius submitted his doctoral thesis on the electrical conductivity of solutions of salts to the University of Uppsala. One of his main topics was the way conductivity varied with the concentration of a salt in solution. Arrhenius's thesis was badly written, and he had made enemies of his professors. He secured only a fourth-class pass, which should have meant that his career was over before it had started. Arrhenius nonetheless sent copies of his thesis to leading chemists, including Van't Hoff and Ostwald. Van't Hoff liked what he read, and Ostwald reacted vigorously and positively, as he had also been working on the electrical conductivity of salt solutions. In 1884 Ostwald published a short paper, giving his own results, and making full and generous acknowledgment of the importance of Arrhenius's results. He also went to Sweden to discuss their work (given that Ostwald was already distinguished, while Arrhenius had nearly failed his Ph.D. oral, this was rather startling). Ostwald then arranged for the University of Riga to offer Arrhenius a position, which as it turned out he was unable to accept. All this was an embarrassment to the University of Uppsala, which gave Arrhenius a traveling fellowship that would keep him out of the way and abroad for years. Arrhenius used the fellowship to work with leading figures, including Van't Hoff and Ostwald.

In 1887 Ostwald published Arrhenius's paper on electrolytic dissociation in his new *Journal for Physical Chemistry.* Arrhenius proposed a dynamic equilibrium between undissociated molecules and their ions in solution. Van't Hoff and Ostwald both incorporated Arrhenius's ideas into their own work on solutions, giving Arrhenius full credit. In the following year, Ostwald published his own major paper on the dilution law. This was a general mathematical law showing the connection between the concentration of an electrolyte and its molecular conductivity.* The work of Ostwald, Arrhenius, and Van't Hoff on solutions, electrolytes, and thermodynamics showed a remarkable convergence, and the trio became known as the three *ionists.* They not only were powerful figures in the early days of physical chemistry as a discipline; they also changed the balance between experiment and theory—ideas were now developed and then tested in the laboratory, rather than worked out in the laboratory before being codified in theory. And the three researchers emphatically made chemistry a mathematical science.

We cannot leave electrochemistry and the thermodynamics of solutions without acknowledging the important work done by Walther Nernst (1884–1941) on the properties of ions in solution. He obtained his Ph.D. in 1887, was introduced to Ostwald by Arrhenius, and became Ostwald's assistant at Leipzig. In 1891 he became professor of physical chemistry at Leipzig. In the 1880s and 1890s he took electrochemistry from its undeveloped state to one where its thermodynamic aspects were thoroughly developed. In 1894 the *Journal for Electrochemistry* (*Zeitschrift für Elektrochemie*) was founded, again in Leipzig, and Nernst's work provided its editors with support for their view that

---

*Molecular conductivity is more often known today as molar conductivity. "It can be visualized as the conductivity between two parallel plates a fixed distance (e.g. 1 cm.) apart, of such an area that one mole of the solute is present between them" (Laidler, *The World of Physical Chemistry,* 209). A mole is the molecular weight of a substance in grams. Thus if the molecular weight is x, the molar weight is x grams. It soon became apparent that Ostwald's equation, predicting decreasing molecular conductivity with increasing concentration in the solution, did not work for all ordinary salts or for some acids and bases. We now know that such substances dissociate completely into ions in solution, so there is no changing equilibrium between ions and undissociated molecules.

electrochemistry included all of physical chemistry. In 1904, the title of the journal was changed to the *Journal for Electrochemistry and Applied Physical Chemistry* (*Zeitschrift für Elektrochemie und angewandte physikalische Chemie*).

How Fast, and How Far?

Physical chemistry, in the hands of the ionists, looked above all at questions of equilibrium and yield, using chemical thermodynamics. But physical chemistry, as the ionists were well aware, was more than the study of equilibrium and yield, and more too than the study of electrolysis. *Chemical kinetics,* the study of reaction rates, of the speed with which reactions take place, also has an important and early place in the history of physical chemistry. And, just as electrochemistry began well before the clear assertion of the discipline of physical chemistry, so too did chemical kinetics.

The first research in which a mathematical approach was taken to reaction rates was when the German chemist Ludwig Ferdinand Wilhelmy (1812–64) studied the *inversion* of sucrose. When an optically active solution of cane sugar is left to itself, or warmed gently with dilute acids, it decomposes into two substances, and in the process, the direction in which polarized light passing through the system is rotated changes from right to left. This reversal of optical rotatory power, and the accompanying decomposition of the sugar, is known as inversion. Wilhelmy used a polarimeter (a device which measures optical rotation) to track the progress of the reaction at different concentrations of acid. He found that at any instant, the rate of change of the sugar concentration was proportional to the concentration of both the sugar and the acid. He used calculus to set up an equation that enabled him to calculate the rate with changing sugar concentration, and he found that theory and experiment matched. In 1884 Ostwald drew attention to Wilhelmy's neglected work, reprinting the paper in his series of "classics."

Other chemists carried out kinetic studies in France and Norway in the 1860s. The next major step, however, occurred through the collaboration of a Scottish mathematician, William Esson (1839–1916), and his English chemical colleague, Augustus Vernon Harcourt (1834–1919). Esson worked out the mathematics for calculating rates of reaction in both first-order reactions, where the rate is proportional to the concentration of a single substance, and second-order reactions, where the rate is proportional to the concentration of two substances. He also worked out the equation for consecutive reactions, where the product of one first-order reaction itself undergoes another first-order reaction. Harcourt conducted extremely careful quantitative laboratory studies of reaction rates, and Esson used his mathematics to analyze the results.

In the final years of their long collaboration, Harcourt and Esson investigated the way in which reaction rates depended upon temperature. When Van't Hoff wrote *Studies in Chemical Dynamics* (1884), embracing chemical kinetics and thermodynamics, he was able to build on the work of Harcourt and Esson. He showed more fully how temperature influences reaction rates, and Arrhenius in turn took his work further. The history of chemical kinetics in its early years is one of cooperative and complementary work, with little of the feuding that characterizes the development of many areas of science.

13 The Nature of the Chemical Bond

Chemical affinity is a concept that has been around for a long time but has kept changing. For centuries, it signified the attraction of similar substances for one another. Then it shifted to mean the attraction of opposite or unlike substances. In the seventeenth century, chemists invoked the fit of geometrical shapes and then Newtonian attraction to account for the way that different chemical substances had diverse but distinct attractions for one another. In the early eighteenth century, especially in France, affinity became a central organizing theory for classifying chemical substances and their reactions, but many chemists were reluctant to allow the concept of affinity any explanatory power.

Around 1800, Berthollet in France revived the Newtonian model for affinity, seeing chemical reactions as the result of the interplay of forces of attraction between atoms and molecules. In the ensuing decades, Berzelius in Sweden and Davy and Faraday in England were among those chemists who thought of affinity as the attraction of substances characterized by their electrical natures; affinity was the attraction of opposite electricities, positive and negative. Then, in midcentury, affinity all but vanished as a topic of chemical discourse. Atomic and molecular weights and the classification of elements and compounds were more urgent subjects for most chemists. Valence theory and structural chemistry in the third quarter of the century had implications for theories about the extent and direction of chemical links or bonds between atoms, but they had nothing to say about what constituted those links. Only toward the end of the century did affinity come back in a significant way, when Van't Hoff and others made it an important part of chemical thermodynamics, a key instrument in the advancement of physical chemistry.

There were, however, two other areas of chemistry and physics that laid the foundations of a new theory of affinity, based on a developing understanding of atomic structure and the related fields of spectroscopy and quantum theory. Once we get into these waters, any distinction between chemistry and physics is often artificial. As we saw in the last chapter, Nobel Prizes in chemistry sometimes went to scientists who thought that they were physicists. For now, we shall not worry about labels, but simply show how a series of initially unrelated developments came together. The result was a new understanding of the nature of the chemical bond, but the beginnings were far from the preoccupations of most chemists.

Sunlight, Starlight, and the Colors of Flames

When sunlight passes through a glass prism, it forms a spectrum, showing all the colors of the rainbow. This was the starting point for Isaac Newton's revolutionary account of color. For Newton, white light was not simple, but rather was made up of different colors—an improbable state of affairs that ran dead against the theories of light from the time of the ancient Greeks until the final decades of the seventeenth century. Newton described the colors of the spectrum of solar light with his customary precision. However, because he and his eighteenth-century successors did not use sufficiently narrow slits in the screens that served to isolate a beam of light, they were unable to see that there are also dark lines in the solar spectrum. These lines were observed in the 1800s, but they were first accurately recorded and studied in 1817 by Joseph Fraunhofer (1787–1826).

Fraunhofer was a lens maker, who, after noticing a pair of yellow lines in the spectrum of a flame, went on to look at the spectrum of sunlight and discovered several dark lines there. He later examined the spectrum of starlight and discovered that the lines in the spectrum of light from different sources were often different from one another. He concluded, from meticulous observations and experiments, that the lines were characteristic of their sources, of the light itself and of the substances through which that light passed. Chemists had known for centuries that the flames emitted on heating different substances differed in color. Copper salts, for example, colored a flame green. This was a working assumption when chemists, mineralogists, and metallurgists used blowpipes to identify different substances. It was reasonable for some chemists in the nineteenth century to suggest that Fraunhofer's spectra, with their colored and dark lines, might be useful in chemical analysis.

Robert Bunsen and Gustav Kirchhoff, whom we encountered in passing in the previous chapter, took a major step in this direction in the middle of the nineteenth century. Bunsen moved to Heidelberg in 1852, and there he worked on combustion. Kirchhoff joined Bunsen two years later. It was Kirchhoff the physicist who suggested to Bunsen the chemist that a prism would be useful to examine the color of flames as they were affected by different metallic salts.

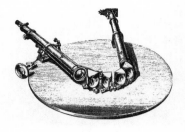

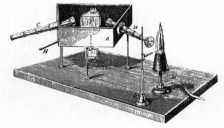

The Spectroscope

The righthand illustration shows the simple and elegant construction of Bunsen and Kirchhoff's first form of spectroscope. A prism is supported within a box (*A*). The light to be analyzed is passed to the prism through the righthand tube (*B*). This tube is known as a *collimator*, which would be more intelligibly named if it were called a *collineator*, since its job is to line up or align the light source with the prism. At its outer end, it has a fine slit, which can be adjusted by a screw. The other end holds a lens, which collects the rays coming from the slit and makes them parallel before they reach the prism. The light passes through the prism, which pro- duces a spectrum that is viewed through the tube on the left, which is a low-powered telescope.

The light can be passed through several prisms, as is shown in the illustration at left. That gives wonderful separation (devi- ation and dispersion) of the different wavelengths, which is easy to read but also results in a weakening of the light, since the light is spread out by each successive refraction and much is lost by reflection. Such an instrument is therefore useful only for analyzing light from very bright sources, such as the sun or a carbon arc lamp.

■ J. Norman Lockyer, *Solar Physics* (Lon- don, 1874), 158, figures 47 and 48.

Their collaboration resulted in the production of the spectroscope, with a moveable prism and fixed lenses.

The spectroscope, with a scale added, made it possible to record the brightly colored lines and bands as well as the dark lines in a spectrum with ease and precision. Bunsen and Kirchhoff used their instrument to good ef- fect, discovering new metallic elements and showing that some dark lines in the solar spectrum were in the same place as bright lines in flame spectra. Kirch- hoff argued that an element when heated would emit light that produced a particular set of colored lines, that is, light of particular wavelengths. When light of those wavelengths passed through the vapor of the same element, that element would absorb it. Elements could absorb light of the same wavelength that they emitted when heated. The dark lines that Fraunhofer had observed

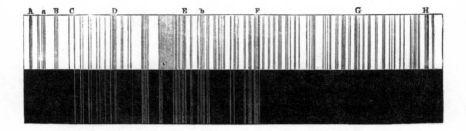

Spectrum of Iron: Coincidence of Bright Lines with Fraunhofer's Dark Lines

Sodium salts, when heated in a flame, give that flame a bright yellow color, and this color matches the two brightest lines in the emission spectrum of sodium. Sodium is found widely in nature, and lots of substances produce these lines. Fraunhofer had found that the spectrum of a candle flame contained two bright lines precisely corresponding to two dark lines, known as the D lines, in the emission spectrum of the sun.

Kirchhoff and Bunsen concluded, after extensive joint observations, that "the dark lines of the solar spectrum which are not evoked by the atmosphere of the earth exist in consequence of the presence, in the incandescent atmosphere of the sun, of those substances which in the spectrum of a flame produce bright lines at the same place" (quoted in Lockyer, *Solar Physics*, 193). The spectrum of sodium is a remarkably simple one; most elements have more complex spectra. Kirchhoff went on to examine the other lines of the solar spectrum (a labor that cost him one of his eyes), until he had a sufficiently complete spectrum to determine which metals were, and which were not, in the solar atmosphere.

This figure shows the match between the bright lines in the spectrum of iron with some of the Fraunhofer lines, proving the presence of iron in the solar atmosphere.

■ J. Norman Lockyer, *Solar Physics* (London, 1874), 197, figure 82.

were characteristic of absorption spectra and were to be expected in the same places in the spectrum that colored lines occupied in emission spectra from the same element. Different elements produced different spectra. In 1861 Bunsen and Kirchhoff had discovered cesium from the blue lines and rubidium from the red lines in their spectra. In the same year William Crookes, a brilliant but eccentric English chemist who believed in spiritualism and proposed an evolutionary theory of matter to account for the chemical elements, discovered another new metal, thallium, from the vivid green line in its emission spectrum.

It was not long before chemists used the spectroscope to analyze light from the sun and stars, showing what elements were present in them, and even detecting new ones. Helium, an inert gas, was discovered from a new and unusual line in the solar spectrum before it was found on earth. Clearly, spectroscopy was a very useful tool for chemists. With the application of photography, spectroscopy became even more efficient, and the way was open to a thorough recording, observation, and analysis of the solar, stellar, and terrestrial spectra.

Chemists were fascinated by periodic regularities in the properties of elements. Physicists were and are more prone than chemists to seek regularity and pattern in numbers, and the distribution of lines in spectra became a subject of intense mathematical inquiry, and often resulted in intense frustration. But there were isolated successes too. In 1885 Johann Balmer, a Swiss schoolteacher of mathematics, found that the wavelengths of four lines in the spectrum of hydrogen could be precisely represented by a simple equation if one of the variables in the equation was successively given the values 3, 4, 5, and 6. He predicted a fifth line, which was later observed. Other series of lines in the hydrogen spectrum were later shown to correspond to a generalized form of the same equation. But no physical or chemical theory of the late nineteenth century managed to explain how these regularities arose. Insights into the structure of the atom and the behavior of one subatomic particle, the electron, were to provide the key to the riddle of the spectrum.

The word *atom* comes from a Greek root meaning "indivisible." Chemical atoms, as John Dalton presented them, were indivisible. But other chemists were not so sure about that. The striking number of elements whose atomic weights were close to whole numbers could not be accidental. And so Prout, among others, had hypothesized that atoms were divisible. Later in the century, the identification of allotropy, different forms of the same element with very different physical properties, was another pointer to the same conclusion, that atoms were not the ultimate particles but were in fact divisible into smaller and simpler units. But not all chemists accepted this reasoning. Some, like Michael Faraday, were even reluctant to adopt chemical atomism. Faraday, in his electrolytic researches in the 1830s, had shown that there was a relation of constant proportionality between the weight of a substance deposited in electrolysis and the chemical equivalent weight of the same substance. If Faraday had wanted to, he could have argued from this result, and from the idea that chemical elements consisted of atoms or least particles of constant weight, to the hypothesis that electricity also consisted of particles characterized by their charge. In other words, Faraday could have argued from the existence of chemical atoms to the existence of atoms of electricity. But he did not like the atomic hypothesis in chemistry, and so he did not explore this line of thinking any further.

In 1881 the extraordinary German natural philosopher Hermann von Helmholtz (1821–94) gave a lecture in the series named for Faraday at the Royal Institution of Great Britain. In that lecture, he argued that since there were atoms, Faraday's laws of electrolysis meant that ions, charged atoms or groups of atoms, carried a constant charge. That in turn meant that Davy and Berzelius had been right and that chemical affinity, the cause of chemical bonding, was electrical in nature. So far so good. But although that could be seen as implying that electricity was itself atomic, such a conclusion had to remain theoretical until such time as someone found experimental evidence for atoms of electricity. That time soon came.

From the 1850s on, physicists had studied electric currents in tubes from which almost all the air had been evacuated. Electricity could pass from the electrode (cathode) at one end of the tube to the electrode (anode) at the other. A glow appeared on the walls of such a tube when electricity passed through it. The glow was attributed to rays from the cathode, or cathode rays. (The phenomenon is familiar to anyone who has ever watched television or used a computer monitor; the screen that we view is part of a cathode ray tube.) William Crookes discovered in 1879 that electricity itself produced the light and that the cathode rays were in fact streams of electrically charged particles.

In 1897, Sir Joseph John Thomson (1856–1940), professor of experimental physics at Cambridge University in England, was able to show that cathode rays were made up of negatively charged particles of constant charge, and he determined the mass and the charge of these particles. Each one weighed roughly 1/2000 of the weight of a hydrogen atom. They were vastly smaller and lighter than any chemical atom, and they appeared to form part of the different kinds of matter, of different chemical elements. Within a couple of years, physicists everywhere were talking about the tiny atoms of electricity. Thomson clearly recognized the importance of his discovery: "We have in the cathode rays matter in a new state, a state in which the subdivision of matter is carried very much further than in the ordinary gaseous state; a state in which all matter—that is, matter derived from different sources such as hydrogen, oxygen, etc.—is of one and the same kind; this matter being the substance from which all the chemical elements are built up."[*] Here was clear recognition that electricity was made up of subatomic particles and that all chemical species were constructed from the same kind of building blocks. Thomson called these atoms of electricity *electrons.* The old indivisible chemical atom was well and truly dead.

[*]J. J. Thomson, "Cathode Rays," *Philosophical Magazine* 44 (1897): 293–316, reprinted in Mary Jo Nye, *Before Big Science* (New York: Twayne, 1996), 162.

Steps Toward the Electron Theory of Valence

Now, on the brink of the twentieth century, Helmholtz's conclusion that chemical affinity was electrical in nature could be made more precise. Chemical affinity had to have something to do with the newly discovered electron, which was a constituent of atoms and ions. In fact, although Thomson rightly deserves the credit for discovering the electron and even more for measuring its mass and charge, at least one chemist had, in speculation, been there ahead of him and applied the idea and the name *electron* to chemical affinity. In 1891 the Irish natural philosopher George J. Stoney had written that "a charge of this amount is associated in the chemical atom with each bond. . . . These charges, which it will be convenient to call *electrons,* cannot be removed from the atom; but they become disguised when atoms chemically unite."*

Stoney was remarkably on target, but without Thomson's empirical and theoretical work, his ideas were trapped in the realm of speculation and had little impact. *After* Thomson's work, chemists could easily accept that electrons held the answer to the puzzle of chemical affinity. But just what did electrons do to bring about chemical combination? Some of the most important answers to this and related questions were to come from Gilbert Newton Lewis (1875–1946). Lewis, a brilliant and ambitious American, with a Ph.D. from Harvard, went in the early 1900s to work with Ostwald and Nernst, key players in electrochemistry, before moving to the Massachusetts Institute of Technology and then to the University of California at Berkeley. Earlier chemists had thought about valence in terms of the number and direction of chemical bonds, in both organic and inorganic chemistry.

Lewis addressed the role of electrons in forming bonds between atoms. As early as 1902, he came up with a sketch, but he did not then publish it; indeed, he sat on it for fourteen years. The sketch shows the elements of the second and third periods in the periodic table and presents the outside of atoms as a system of electrons at the eight corners of a cube. For the second period, Lewis drew cubes for lithium, beryllium, boron, carbon, nitrogen, oxygen, and fluorine. He started with one electron for lithium, adding an electron as he moved across the group, and ending with seven electrons for fluorine. He did not sketch the inert gas neon, with its eight electrons, although it had recently been discovered. For the third period he sketched sodium with one electron, through magnesium, aluminum, silicon, phosphorus, and sulfur, to chlorine with seven electrons. He did not sketch at the end of the period the inert gas with eight electrons, which in this case was argon, although it too had been

*G. J. Stoney, *Transactions of the Royal Society of Dublin,* 2nd ser., 4 (1891): 583.

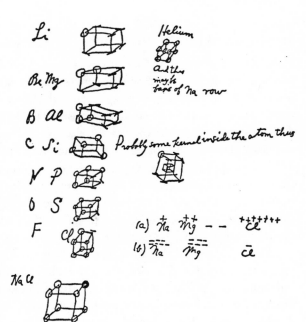

G. N. Lewis's first sketch
of the octet theory of
valence electrons.

■ G. N. Lewis, *Valence
and the Structure of Atoms
and Molecules* (New York:
Chemical Catalog Co.,
1923), 29.

---

discovered in the previous decade. But he did make a sketch of helium, with
eight electrons, and noted that "this may be basis of the Na [sodium] row."*
The inert gases had eight outer electrons and were stable and nonreactive.

Perhaps the most stable state of combination, the one brought about by
the most vigorous affinity, was one where atoms had all eight electrons at the
vertices of their cubes. Atoms, represented by cubes, could increase the number of their electrons to eight by sharing electrons at the corners of the cubes.
In single bonds, one electron was shared between two atoms. For example, in
sodium chloride, sodium shared an electron with chlorine, so that the former
had two electrons, of which one was shared, while the latter had eight electrons, of which one was shared. In double bonds, two electrons were shared.
It is easy to read too much into a sketch when we know how the story worked
out later. But we can, at any rate, say that for Lewis, atoms in any given group
(in the same vertical column in the modern periodic table) had the same number of outside electrons, from one to eight. It is also clear that as early as 1902,
Lewis was considering the notion that chemical combination involved a sharing of outer electrons.

Several developments came one after the other over the next fourteen years

*G. N. Lewis, *Valence and the Structure of Atoms and Molecules* (New York: Chemical Catalog Co., 1923), 29.

to reinforce that way of thinking. The British physicist Sir Joseph Thomson had, around 1900, proposed a model of the atom which became known as the plum pudding model, in which small negatively charged electrons were embedded in a diffuse positive sphere, like currants in a pudding. This did not fit Lewis's cubic geometry, but it did encourage thinking about atoms in terms of electrons and a positive charge. Then came an interpretation of variable valence, a concept that we briefly encountered in looking at Werner's coordination chemistry (Chapter 11).

In 1904, Richard Abegg, a Polish-born physical chemist who worked in Nernst's laboratory, produced a table showing a striking and simple correlation between the variable valencies of elements and the group in the periodic table to which those elements belonged. Elements had a normal or primary valence equal to the number of their group, 1 for sodium, 2 for magnesium, and so on; but they also appeared to have a secondary, or contra-valence. Nitrogen, for example, had a normal or primary valence of 3 in ammonia, but in nitrogen pentachloride, nitrogen's (contra-)valence was 5. The sum of the two valencies was 8, and Abegg noted that this was a general rule.

Group	1	2	3	4	5	6	7
Primary valence	1	2	3	4	5	6	7
Contra-valence	7	6	5	4	3	2	1

Perhaps the sharing of outer electrons would make such combinations as the diatomic oxygen molecule or the diatomic chlorine molecule stable, by giving each atom eight electrons by sharing. But before he could take this idea further, Abegg died in a ballooning accident.

In the following year, 1905, Werner published his great book on coordination chemistry, and this had a powerful influence on Lewis. Werner proposed that in coordination compounds, atoms or groups of atoms surrounded a central atom to form an electrically charged ion or a neutral compound, and the geometrical or structural theory seemed to fit very nicely with Lewis's ideas. All that was needed (and it was a big "all") was a clearer picture of the electrical nature of atoms.

That electrons existed, were crucial constituents of all atoms, and were the key to chemical bonding was widely accepted. In 1899 the New Zealand physicist Ernest Rutherford (1871–1937), one of the towering figures in the history of the atom, discovered positively charged particles that he called *alpha* particles. By 1902, the influential scientific journal *Nature* referred to matter as consisting of "positive and negative electrons." In 1911, Rutherford caused alpha particles to strike a sheet of gold foil. Most of the alpha particles went right through the foil, but some of them were reflected, while still others were scattered at wide angles. Rutherford explained the reflection and scattering as caused by collisions with a heavy positively charged nucleus, which had to be much heavier than electrons. Because most of the alpha particles passed through the foil, the nucleus had to be much smaller than the atom of which it was a part. That suggested that most of an atom was space between the positive nucleus and the negative electrons, which, Rutherford suggested, orbited around the nucleus like planets around the sun. Every atom was like a miniature solar system. Lewis's sketch of 1902 took no account of the *movement* of electrons, but the notion of outer electrons outside a positive nucleus was getting strong support from the physicists.

In the next section we shall see how physicists modified and took Rutherford's model even further, in ways that made brilliant sense of chemical spectroscopy. But that can wait. Lewis was not influenced by these later developments in his version of chemical bonds. He was, however, ready by 1916 to publish the diagram that he had scribbled down in 1902. Atomic shells, he announced, could penetrate one another, so that an electron "may form part of the shell of two different atoms, and cannot be said to belong to either one exclusively."* In 1923, he published his major work on chemical bonding, *Valence and the Structure of Atoms and Molecules.* He argued that when atoms combined, electrons were paired in their shells. They could even, in special cases, move completely from one atom to another. Lewis's theory allowed, in other words, for the formation of ions and for ionic combinations. For example, in sodium chloride, Na^+Cl^-, sodium is envisaged as having given up an electron to chlorine, so that the latter has its eight electrons, while sodium has emptied its outer shell.

Lewis presented his theory as follows: "Two atoms may conform to the rule of eight, or the octet rule, not only by the transfer of electrons from one atom to another, but also by sharing one or more pairs of electrons. The electrons that are held in common by two atoms may be considered to belong to the outer shell of both atoms." Valence was then the number of electron pairs that an atom shared with one or more other atoms. Single bonds corresponded to a sharing of one electron pair, as in the hydrogen molecule, H:H, where each dot represents an electron. In double bonds, two electron pairs were shared, as in the oxygen molecule, :O::O:, while in triple bonds three electron pairs were shared, as in the nitrogen molecule :N:::N:. Lewis had established the notion

*G. N. Lewis, "The Atom and the Molecule," *Journal of the American Chemical Society* (1916): 762–65, quoted in W. H. Brock, *The Norton History of Chemistry* (New York: Norton, 1993), 476.

of electron pairs as the key to covalent bonding, while ionic bonding involved the transfer rather than the sharing of electrons between molecules.

Much of what Lewis had to say about bonds is still useful, but a third line of development that he pretty well ignored has also turned out to be challenging and fruitful for an understanding of the nature of the chemical bond. That line of development goes back to spectroscopy and forward into the extraordinary area of twentieth-century physics known as quantum theory.

From Black Bodies to Orbitals and Bonds

So far, in looking at spectroscopy, we have looked at what was *different* in the spectra of various chemical species. Spectroscopy was a tool of chemical analysis. But *black bodies* when heated to the same temperature all emit the *same* pattern of radiation. A black body is a hypothetical or ideal body that absorbs all the radiation that falls on it and then emits that radiation with a distribution of wavelengths that is independent of the chemical nature of the body. This is precisely the kind of problem that physicists like, something that is unaffected by the complications introduced by chemical diversity.

One promising explanation of radiation was that the atoms of bodies, when heated, oscillated at faster rates at higher temperatures and emitted radiation at the frequency at which they were oscillating. High frequency means low wavelength, and so hotter bodies would emit more radiation at low wavelengths than cooler bodies did. Statistical theories were used to calculate the distribution of oscillation frequencies among the atoms of a body at different temperatures. From this distribution one could calculate the distribution by wavelength of the radiation emitted by the body at any temperature. One could do this by using thermodynamics or by using a combination of mechanics and statistics known, unsurprisingly, as *statistical mechanics.*

Physicists thought that they would be able to predict the emission spectrum of a black body at any temperature. Radiation has energy. If we plot a graph of the energy of emitted radiation against the frequency of that radiation, then late nineteenth-century theory gives us a curve where the energy increases at high frequencies, that is, at the ultraviolet end of the emission spectrum. The trouble with that prediction is that it was falsified by experiments, which clearly showed that the energy decreased at high frequencies. This was a disaster for the theory and has become known as the *ultraviolet catastrophe.* Classical mechanics could not handle emission spectra.

If theory will not predict observed fact, then another approach is needed. Max Planck (1858–1947), a German physicist, set about to provide a mathematical description of the actual distribution of frequencies in the observed spectrum of radiating bodies. He came up with an equation as simple as it was perplexing. His description gave oscillating atoms in the radiating body an energy $E = nh\nu$, where n was a small whole number (1, 2, 3, etc.), h was a constant that became known as Planck's constant, and ν was the fundamental or lowest frequency of oscillation. This meant that the oscillator could vibrate only at frequencies of $h\nu$, $2h\nu$, $3h\nu$, and so on; that is, at twice, three times, four times the fundamental frequency, but not at any intermediate frequency. The energy existed at separated and discontinuous levels; it was *quantized.* It was as if an automobile could travel at 10 mph, 20 mph, 30 mph, but not at any intermediate speed. This was so much at odds with classical notions of continuity in physics that it seemed frankly crazy, and Planck and others took a while to accept the implication of the mathematics.

In 1905, the year in which Albert Einstein (1879–1955) published his great paper on the special theory of relativity, Einstein published another paper, on the photoelectric effect, that had immediate significance for Planck's work. The photoelectric effect is something that we take for granted today, since it is used in photoelectric cells in every automatic camera. Light falls on a metallic detector, knocks electrons out of the detector, and triggers an electric current. The strength of the current gives our camera or us a measure of the intensity of the incoming light, that is, it tells us how bright the light is. But light below a certain frequency, no matter how bright the beam, does not produce any current, whereas light above that frequency, even at low intensity, would produce a current. Here, as in black body radiation, is a problem insoluble within the framework of classical physics. Einstein correlated the threshold frequency for the photoelectric effect with the fundamental frequency in Planck's equation. He argued that light came in packages, or *quanta,* characterized by their energy, and that energy needed to be at least great enough to cause the oscillators in the detector to vibrate at the minimum frequency that would enable it to shed an electron, that is, to produce an electric current.

By now, it was becoming clear that there was a connection between electrons in bodies, the radiant energy emitted by those bodies, and the distribution of that energy in the spectrum. But a more detailed theory with more information was needed. Rutherford had proposed an atom modeled on the solar system, with electrons orbiting around a positive nucleus and a lot of empty space between the electrons and the nucleus. In 1913 the Danish physicist Niels Bohr (1885–1962), who worked with Rutherford for four years and on his return to Copenhagen made Denmark a world center of theoretical physics, published one of the twentieth century's most important papers. He applied Planck's equation and the notion of quantization of energy to Rutherford's planetary atom, which was otherwise open to a crippling objection. The solar system is stable, but Rutherford's atom was not; electrons, as charged bodies attracted to the nucleus, would, according to classical theory, spiral in toward the nucleus, emitting radiation along the way. The result would be the collapse of the atom. Bohr put restrictions, *quantum* restrictions, on the electrons. They could orbit only at energy levels that satisfied Planck's equation, and, within the atom, they could absorb or emit energy only in amounts corresponding to the difference in energy between two levels. With these restrictions in place, Bohr had rescued Rutherford's planetary atom. Quantum restrictions prevented the collapse of electrons into the positive nucleus, and electrons were held in stable orbits about the nucleus, like planets orbiting around the sun.

If that were all that Bohr's quantum restrictions achieved, they would not have been very useful. But, working with a planetary atom subject to the established laws of physics and the new quantum restrictions, physicists and chemists were able to begin to interpret some of the lines in the hydrogen spectrum. These lines were produced when an excited electron—an electron that had been stimulated by the absorption of energy to a higher level than its equilibrium level or to a higher orbit in the planetary model—emitted energy and fell back to a lower level. The difference in energy between the equilibrium level for an electron and the higher level to which the electron had been temporarily excited corresponded to the energy of the radiation absorbed in the excitation. It was also related to the energy of the radiation emitted when the electron fell back to its equilibrium state or level. The energy of radiation is directly proportional to the frequency of that radiation, and so a substance's frequency in the spectrum was directly related to the structure of the atoms that made up that substance, and in particular to the electronic structure of the atoms. But the calculations were far from easy.

The mathematical treatment of the Rutherford-Bohr atom was especially productive in Denmark and Germany. It led directly to *quantum mechanics,* which treated electrons as particles. Electrons, however, like light, were part of electromagnetic radiation, and radiation was generally understood to be a *wave* phenomenon. In 1924, the French physicist Prince Louis de Broglie (1892–1987), influenced by Einstein's work on the photoelectric effect, showed that electrons had both wave and particle aspects. *Wave mechanics,* an alternative approach to quantum physics, was soon developed, based on the wave equation formulated in 1926 by the Austrian-born Erwin Schrödinger (1887–1961). Quantum mechanics and wave mechanics turned out to be complementary, and both were fruitful for an understanding of valence.

The first dramatic chemical advance using quantum mechanics was the calculation of the energy of the hydrogen molecule by two young physicists, Fritz London (1900–1954) and Walter Heitler (1904–81). The quantum mechanics of spectra has never looked back since their work. Gerhard Herzberg (1904–99), working first in Germany and then in Canada (he had left Europe with his Jewish wife to escape the Nazis), began publishing on this topic in 1928, and steadily produced a series of path-breaking books on atomic and molecular spectra.

Meanwhile, thanks to the work of its proponents, wave mechanics was also advancing. The Schrödinger equation contains a *wave function,* represented by the symbol Ψ. The square of that wave function, Ψ^2, indicated the probability of finding an electron in a given location in an atom. A three-dimensional graph based on this probability produced some remarkable images of *electron orbitals.** For the lowest level, $n = 1$ in Planck's equation, the graph indicated that electrons were most likely to be found at the center of a sphere or spherical cloud centered on the nucleus. This became known as the *s* orbital. For $n = 2$, the distribution corresponded to a kind of dumbbell, or rather to a set of three dumbbells along the three spatial axes; these became known as the *p* orbitals. For higher numbers, the shapes or orbitals became less simple and more fantastic. They did, however, suggest a way to visualize the chemical bond, based on the most probable distribution of electrons within the space of atoms. The versatile and outspoken American chemist Linus Pauling (1901–94) took the lead in this interpretation.

Linus Pauling and Electrons in Atomic Space

Pauling was born in the United States and wrote his Ph.D. dissertation in chemical engineering at the California Institute of Technology, where he spent most of the rest of his career. But before settling in as a faculty member at Caltech, he won a Guggenheim Fellowship, visited Europe in 1926, and made the rounds of some of the major figures in quantum physics and quantum chemistry. He visited Bohr in Copenhagen and Schrödinger in Zurich. He met Fritz London in Switzerland and Walter Heitler in Germany. He was taken by surprise when London and Heitler published their account of the hydrogen molecule, but he quickly saw its importance.

Pauling argued that the distribution of the electron cloud, plotted as the probability of finding an electron at different points in space, showed where chemical bonds, valence bonds, were most likely to be formed. When orbitals

---

*The three commonest orbitals are called *s, p,* and *d* orbitals. These letters correspond to the *s*harp, *p*rincipal, and *d*iffuse series of lines identified in early chemical spectroscopy by two Cambridge chemists, James Dewar (1842–1923) and George Liveing (1827–1924).

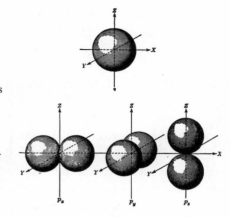

The relative magnitudes of *s* and *p* orbitals in dependence on angle.

■ From Linus Pauling, *The Nature of the Chemical Bond,* 3rd ed. (Ithaca, N.Y.: Cornell University Press, 1960), 109 (© 1960 Cornell University; used by permission of the publisher).

of two atoms overlapped, a bond could be formed. Overlapping *s* orbitals produced a *sigma* (σ) bond; overlapping *p* orbitals produced a *pi* (π) bond.

Quantum physicists in Europe had worked out rules for the distribution of electrons. Lower energy orbitals or shells were filled first, and *s* and *p* orbitals could each contain a maximum of two electrons. The oxygen atom, for example, has two electrons in its *s* orbital and four in its three *p* orbitals. Two of the latter four electrons are paired in one of the *p* orbitals, while the other two electrons appear one each in the other two *p* orbitals. That means the oxygen atom has six electrons and can accept two more to fill up its *p* orbitals. The hydrogen atom has only one electron, in its *s* orbital, and can accept another electron to fill that orbital. In water, two hydrogen atoms can each share their electron with one of the single electrons in oxygen's *p* orbitals. In other words, Pauling created a detailed model for the electron pair bond proposed by Lewis.

In 1931, Pauling managed to work out the bonding for the tetrahedral carbon atom. If one simply allocates electrons to distinct orbitals, then carbon should have three bonds at right angles, corresponding to three *p* orbitals, and a fourth, somewhat weaker bond, based on the *s* orbital, in "some arbitrary direction. This is, of course, not so; and instead, it is found on quantum mechanical study of the problem that *the four bonds of carbon are equivalent and are directed toward the corners of a regular tetrahedron,* as had been inferred from the facts of organic chemistry."* Quantum mechanical equations and laboratory chemistry both point to the equivalence of the four carbon-hydrogen bonds in methane, based on a tetrahedral carbon atom. The three *p* orbitals and the one *s* orbital form four identical hybrid bonds. The term *hybrid* is an

*Linus Pauling, *The Nature of the Chemical Bond,* 3rd ed. (Ithaca, N.Y.: Cornell University Press, 1960), 111.

apt metaphor from zoology. Pauling solved the problem of double bonds, too, for example in ethylene, $H_2-C=C-H_2$, where quantum mechanical calculation for the double bond corresponds to molecular orbital s-p hybrid bonds.

Chemists had moved, in half a century, from first thinking about atoms in space to thinking about the electrical nature of subatomic particles (essentially a part of physics), and then had gone on to consider the role of the electron in chemical bonding. They had managed to explore the energy of that bonding, the ways in which that energy could change, and how those changes shed light on chemical spectroscopy. They had also reached the point where it was productive to think in new ways about the geometry of the chemical bond, the place of electrons in space. They had brought laboratory science and theoretical chemistry into close interdependence. The story is not only rich and complex in matters of scientific theory and practice, it also involves more people in more countries in a shorter period of time than previous major developments we have encountered in the history of chemistry. The growth of the scientific community, the expansion of scientific institutions, and the proliferation of tools for scientific communication, all contributed to the speed with which difficult science was accomplished.

Despite extraordinary achievement, many problems remained, and still remain. We do not have a single theory of the chemical bond that meets all our needs. The equations for the bonds in even the simplest molecules are difficult to solve, although Herzberg was triumphantly successful in his work on atomic and molecular spectra. But the quantum mechanical equations for complex molecules are still too difficult for us to solve in detail. An insightful comment in 1972, prompted by debates about rival theories for interpreting chemical bonding, remains valid even today:

> In physics it is possible to develop a simple and detailed model to explain certain classes of phenomena, but chemistry is too complex to be fully explained by such simple theories. To explain chemical phenomena at the present time, one needs several good models. But these "good" models are more flagrantly models, i.e. they explain only a selection of data, and hence the need for several models. Depending upon the symbolic apparatus used, different truths emerge.*

This chapter has been harder to place than its predecessors. Some of the scientists who contributed to the development of an understanding of the chemical bond were not thinking at all about chemistry, but only about

---

*Quoted in Brock, *The Norton History of Chemistry,* 505, from Robert Paradowski, "The Structural Chemistry of Linus Pauling" (Ph.D. diss., University of Wisconsin, 1972).

physics. Some of them were experimentalists; others were theoreticians whose work was almost entirely in the realm of mathematics. Some of them received Nobel Prizes in chemistry, although they regarded their work as strictly physics. Chemistry, not for the first time, was becoming blurred at the edges, invading and being invaded by physics. As we will see in the concluding chapter, chemistry is still very much a science in progress.

Chemistry and physics, as we have seen, have blended into one another in several areas. It has from time to time been fashionable to say that chemistry can be reduced to physics or that biology can be reduced to chemistry. That implies that the methods and concepts of physics can explain all chemistry and that the methods and concepts of chemistry can explain all biology. Such a way of thinking is called *reductionism,* and it has not proved very fruitful. Scientists use methods and tools that work successfully for them. If chemical tools can explain something about medicine or biology, that is all to the good. But it comes down to a question of complexity and of what kinds of explanation work best in tackling a particular problem. To argue for reductionism begs the question of complexity.

In this, the concluding chapter of our journey through the history of chemistry, we shall look at topics where chemical methods or ideas have proved useful, but not worry further about drawing a line around the science. Nor shall we worry about drawing a line between pure and applied science. Many industries employ chemists to do pure research, in the reasonable expectation that some of it will prove useful. Most chemists are employed in applied science; that is the aspect of chemistry that has had the greatest effect on our environment and on us. In the past one hundred and fifty years, chemical synthesis has become ever more powerful, and it is fair to say that chemistry is the only science that now builds or creates much of what it goes on to study, from artificial elements to the latest plastics and the most powerful pharmaceutical chemicals, from fertilizers to microchips. Chemists have been enormously successful in their explorations, and the results of their work have transformed the world in which we live and work.

Chemistry is what chemists do, and the way that chemists define their science in relation to other sciences is in the end a piece of territorial assertion, a reflection of the norms accepted within the chemical community. It is more a sociological and cultural phenomenon than a reflection of boundaries corresponding to fundamental divisions in nature. Now, in the twenty-first century, what chemists do, and how this affects every one of us in our daily lives, has become more far-reaching than in any previous century. It is not possible, in brief compass, to consider everything that chemists are doing today. Instead, let us pick up some of the themes already developed and see how they have pushed limits, and often helped to define and point to a resolution of a host of problems.

Making New Elements: The Chemistry of the Stars, and the Factories of Radiochemistry

One theme that has constantly resurfaced in the history of chemistry and throughout this book is thinking about what chemical elements are and classifying them. Lavoisier, in the chemical revolution of the late eighteenth century, provided a practical rule for treating elements and compounds. If chemists could not decompose a substance, they had to regard it as an element; if they could decompose it, they had thereby shown that it was a compound. A substance regarded as elementary would be reclassified as compound if chemists later succeeded in decomposing it. Then John Dalton at the beginning of the nineteenth century gave the first clear articulation of chemical atomism, where different atoms were identified with different chemical species. Dalton brought atoms and elements into a one-to-one correspondence.

Spectroscopy was one of the tools developed in the nineteenth century that enabled chemists to identify elements in trace amounts, and also to do what some philosophers had said was impossible: to determine what elements were present in the sun and stars. Physicists in the latter half of the nineteenth century regarded the sun as a giant furnace, in which chemical processes like those on earth generated vast quantities of heat and light. They also calculated, using chemical thermodynamics, how long the sun could continue to burn; and they came up with a figure much less than the vast ages proposed by the proponents of evolutionary biology.

By the early twentieth century, chemists and physicists recognized that the atoms of which chemical elements are composed are themselves made up of *electrons* and *protons,* of electrically negative and positive subatomic particles that were the universal constituents of all chemical elements. Sir Joseph Thomson had discovered the electron in 1897. Ernest Rutherford postulated the existence of a positive nucleus in atoms in 1911, and he used this in developing his planetary model of the atom, with a positive center and orbiting electrons. He discovered the proton in 1919, in experiments on the disintegration of atomic nuclei. Much later, in 1932, the British physicist James Chadwick (1891–1974) discovered a third subatomic particle, the electrically neutral neutron.

The development of nuclear physics, relating the building blocks of matter and energy, showed that the sun could burn as a nuclear furnace for the long ages of time required by evolution. The heart of the sun was a furnace for nuclear chemistry, not for chemical combustion. Then astronomers and cosmologists began to work out ways in which chemical elements were produced in the nuclear furnaces of the sun and stars.

Meanwhile, chemists learned to interpret the periodic table as showing elements arranged according to their atomic numbers, that is, according to the number of protons in their nuclei, rather than according to their atomic weights (the sum of the weights of the protons and neutrons in their nuclei).* Hydrogen has one proton in its nucleus, helium two, and so on. This understanding was the achievement of the British physicist Henry Moseley (1887–1915), who used X-rays to investigate the positive charge on atomic nuclei. His work resulted in important improvements to Mendeléev's periodic table, and consequently to an improved classification of the elements, including the perplexing group of metals known as the lanthanides, or rare-earth elements, with fifty-eight to seventy-one protons in their nuclei.

Because like charges repel one another, the protons in a nucleus exert repulsive force on one another. This repulsive force is balanced by a strong short-range force (the *strong interaction*) that holds them together, but the repulsive forces put limits on the number of protons in a nucleus and make the heaviest nuclei, with many protons, unstable. They break down in a process of radioactive decay, a spontaneous process in which the nucleus splits and emits charged particles and sometimes gamma rays. The Polish-born French chemist Marie Sklodowska Curie (1867–1934) and her husband, Pierre (1859–1906), discovered radium in 1898; together with the French physicist Antoine Henri Becquerel (1852–1908), they were awarded a Nobel Prize for their discovery of radioactivity. Uranium, from which the Curies had isolated radium, had been discovered long before, and Becquerel had shown in 1896 that it was radioactive. Uranium has the highest atomic number and the highest atomic weight of the naturally occurring chemical elements on earth.

Both chemists and physicists subsequently used powerful instruments to produce artificial radioactive elements. Plutonium was first made in 1941 by the American chemist Glenn Seaborg (1912–99) using the cyclotron at the Uni-

*As pointed out in the discussion of isotopes in Chapter 9, the atomic weight of an atom is the sum of the weights of neutrons, protons, *and electrons* in that atom; for all but the most precise purposes, the electrons, which are very light, can be ignored. Where an element has more than one isotope, that is, has different forms containing different numbers of neutrons in their nuclei, the atomic weight will reflect this fact.

versity of California at Berkeley. It is the most frequently mentioned of these new elements, because of its importance for the production of nuclear weapons. In 1951, Seaborg was awarded the Nobel Prize in chemistry for his work in the discovery of plutonium and nine other artificial elements.* Some radioactive elements are used in building atom, hydrogen, and neutron bombs. Some are used in the peaceful production of atomic energy, a field of vast potential that also poses serious problems.

Radioactive substances also have life-saving uses. A radioactive form of cobalt is extensively used in radiation therapy for cancer patients. The treatment was first developed by Harold Johns (1915–) in Canada, where he pioneered cobalt therapy units at the University of Saskatchewan. One of the artificially made elements, Americium (atomic number 95, i.e., with 95 protons in its atomic nucleus), is another life-saving radioactive element. As it decays, it emits alpha particles, which strip electrons from surrounding gas molecules; ionized air conducts electricity much better than air containing smoke particles, and the reduction in conductivity produced by smoke is what triggers the alarm in smoke detectors.

At the time of writing, the latest artificial element to be created (in Berkeley, California, in 1999) is one with the unlikely temporary name Ununoctium (1-1-8-ium), because it has 118 protons. It and its near neighbors in the periodic table are so unstable and short-lived that it is hard to imagine they will prove to have any useful applications. But one never knows.

Electrons in Chemical Action

Spectroscopy and the electron theory of valence provided valuable support for one another. Together, they took our understanding of the nature of chemical elements to a new level, where chemical behavior and chemical structure could both be interpreted in terms of the number and disposition of electrons in the atoms of any given element. At least, the simplified model of atomic orbitals brilliantly developed by Linus Pauling enabled him to explain and predict a great deal of chemistry, in terms of bonds and structures.

A simplified model was necessary, because the exact calculation of all the orbitals for an atom with several shells of electrons is impossible; there is no analytical solution for the Schrödinger equation for atoms with more than one electron. By concentrating on the outer electrons only, and by using the orbitals of these electrons to provide graphic images of electron density (i.e., of where electrons are most likely to be found), Pauling generated an intuitively

*Among the artificial radioactive elements of which Seaborg was discoverer or co-discoverer are Americium, Curium, Berkelium, Californium, and Seaborgium.

Periodic Table

There is nothing like the development of the periodic table through time to give one a sense of the pace of chemical discovery. Lavoisier listed close to thirty elements, and this number more than doubled when Mendeléev invented the periodic table. Since then, we have added the lanthanides and actinides, as well as a stream of artificial radioactive elements.

The table shown here, taken from a book published in 1998, goes as far as the element provisionally named Ununibiium (1-1-2-ium, since its atomic number is 112), created by an international team led by Peter Armbruster in 1996. It had to be created rather than discovered, because all ele-

ments with an atomic number greater than 92, that is, all elements with more protons in their nucleus than in the nucleus of uranium, are too unstable to exist naturally. They are all man-made, and the heaviest of them have very short lives indeed, and, in some cases, only a handful of atoms have ever been made.

Today's periodic table contains about forty-five more elements than Mendeléev's first table (see Chapter 9). As of 1999, we had advanced to Ununoctium as the element with the largest atomic number, 118.

■   From Albert Stwertka, *A Guide to the Elements,* rev. ed. (New York: Oxford University Press, 1998), 6.

---

convincing picture of combination between atoms as the overlap of orbitals. In some cases, it was necessary for Pauling to consider the bonds in a molecule as intermediates or hybrids between two or more graphical formulas, none of which singly could represent the actual distribution of electrons and bonds. In the benzene molecule, for example, where all six carbon atoms in the benzene ring are chemically equivalent and each hydrogen atom is bonded to the carbon atoms in exactly the same way, Pauling proposed a hybrid, where *resonance* among five possible structures produces the symmetry in bonding properties that is implied by experiment. Resonance involves a kind of mixture of bonds. The *resonance hybrid* makes the benzene molecule more stable than any of the five formulas.

While Pauling was elaborating his valence bond model, another American, Robert S. Mulliken (1896–1986), was working out a different approach based on *molecular orbitals*. Instead of considering the electron as situated in a single atom or in a bond between two atoms, Mulliken treated the electron as if it were distributed over the whole molecule. This approach, pursued on both sides of the Atlantic, became the dominant one after the end of World War II in 1945. For complex molecules, supporters of the molecular orbital approach argued that their approach was simpler in giving a picture of electron distribution; and certainly, visualizing a benzene ring with a symmetrical electron distribution is simpler than thinking about resonance among five different bond structures.

One of the most striking recent developments in instrumentation that depends on quantum theory and wave mechanics is that of the *scanning tunneling microscope (STM)*. Electrons have wavelike properties that allow them to "tunnel" into regions of space that are permitted by wave mechanics but forbidden by the rules of classical physics. This tunneling produces a small electric current. One can apply this principle to map the surface of conducting substances with astonishing accuracy. Indeed, it is now possible to map the distribution of individual atoms on a surface, using the STM. The probability of finding electrons in regions forbidden by classical physics decreases sharply with the distance from the surface under investigation. The STM uses a sharp tungsten probe to scan the surface of a metal or semi-metal by applying a small current between the probe and the metal surface and measuring variations in the tunneling current. Gerd Binnig (1947–) and Heinrich Rohrer (1933–) invented this instrument and procedure in Switzerland in 1981, for which they received the Nobel Prize for physics in 1986. The surfaces of silicon have been carefully explored because of their importance in microchips. We know just how the atoms are arranged. It is now also possible to use the STM to manipulate small numbers of atoms on a conducting surface; the implications for microminiaturization of computers, among other machines, are exciting.

At the same time that some chemists were developing chemical spectroscopy, others were developing the field of *chemical kinetics,* the study of the rates at

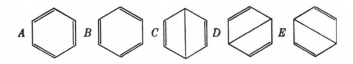

Formulas for Benzene

Structures *A* and *B* are the standard representations of benzene, and the bonds are the way that Kekulé described them. Structures *C, D,* and *E* are the brainchildren of the British-American chemist Michael James Stuart Dewar, who was born in India in 1918.

Kekulé's forms have the virtue that each carbon atom in the benzene ring or hexagon has the same kind of bonding to its neighbors as the other five atoms in the ring. Dewar's forms, in contrast, have different bonding for different carbon atoms, so that four carbon atoms have a single and a double bond to other carbon atoms, while two of the atoms in the ring have three single bonds to other carbon atoms (and one of these bonds is between carbon atoms at opposite sides of the benzene ring).

Unlikely though it may seem, a stable Dewar benzene was synthesized in 1963. But all five forms of benzene contribute to produce a truly symmetrical benzene molecule; the resonance calculations are very complicated.

■ Linus Pauling, *The Nature of the Chemical Bond,* 3rd ed. (Ithaca, N.Y.: Cornell University Press, 1960), 204.

---

which chemical reactions take place. This formed an important part of the research program of Van't Hoff and other early physical chemists. More recently, chemists have begun to explore the events that take place during a chemical reaction. What happens to the atoms and molecules during a reaction? This branch of chemistry is now known as *chemical dynamics.**

A particularly fruitful way into the study of chemical dynamics is through the study of *photochemistry,* the set of reactions caused by the absorption or emission of light. Chemical bonds can be broken by the absorption of energy in the form of light. Thanks to quantum theory, we now have a far deeper understanding of photochemical reactions than was previously possible. And, thanks to some extraordinary developments in instrumentation and laboratory technique in recent decades, we now have a good deal of empirical knowledge of chemical events that were previously completely beyond the reach of experiment.

Individual chemical events, the reactions between atoms, occur rapidly,

*That terminology can be confusing, because what we call chemical kinetics was called "chemical dynamics" by Van't Hoff; not for the first time, the meaning of a technical term has changed with the years.

sometimes unbelievably rapidly. The challenge was to produce a short but intense flash of light that would cause the reaction, and then to develop means for investigating what happened immediately after the flash. The technique that has proved most useful was pioneered in England by George Porter (1920–) and Ronald Norrish (1897–1978) around 1950 and is known as *flash photolysis*. Short intense flashes were followed by spectroscopic analysis. Porter's earliest flashes lasted about a thousandth of a second (10^{-3} seconds). By 1966, Porter had used a laser to produce flashes of around one nanosecond, one thousand-millionth of a second (10^{-9} seconds). That was short enough to study just about any chemical process. Physicists eager to study briefer, purely physical changes using laser spectroscopy have since produced flashes that are one femtosecond, that is, one millionth of a nanosecond (10^{-15} seconds) in duration. This is the timescale on which chemical bonds are formed or broken. The Nobel Prize for chemistry in 1999 was awarded to Ahmed Zewail (1946–) for his work on the transition states of chemical reactions using femtosecond spectroscopy. As a result of such work, we now know far more than before about all kinds of reactions.

Wonderful Structures: Buckyball and Double Helix

For small atoms and simple molecules, we can calculate the electron distribution and make sense of the spectra. We can also use theory to tell us what the angles between bonds and the structure of molecules will be. We can invent new compounds by considering electron distribution and valence bonds or molecular orbitals, and then set about producing those compounds. This is an area where chemists have lately been stunningly successful. New compounds are being synthesized every day at an accelerating rate. In 1965 there were less than half a million compounds known across the whole of chemistry. Today, the figure is more than eight million.

Not all of these compounds, in fact very few of them, have been invented by using quantum mechanics. The calculations are too cumbersome and too complicated for us to work out structures for complex molecules containing many atoms; in fact, the calculations are too complicated for all but the simplest molecules. Chemists use different models for different degrees of complexity. And they usually synthesize a molecule first, and determine its structure afterward.

In the late nineteenth century, Van't Hoff and Le Bel announced the tetrahedral bonding of carbon atoms. This provided an explanation of the existence of enantiomers. But whereas processes in living nature often produced either left- or right-handed molecules, chemists in the laboratory were at first able to produce only mixtures of the two forms, which rotated the plane of polarized light in opposite directions. If chemists wanted to work with just one enantiomer, they had various choices, including obtaining it from natural sources or crystallizing a mixture of the two forms and then separating them physically.* In recent years, a technique called *chiral catalysis* has been studied and applied with great success.

Catalysis is a term invented by Berzelius to describe a process in which a substance facilitates a particular chemical reaction, without itself being consumed in the reaction. *Chirality* is a term first used by Lord Kelvin in 1894 to describe any figure that did not have a mirror-plane of symmetry; optically active organic compounds are thus *chiral.* In chiral catalysis, we usually start with materials that are not themselves chiral. The chirality of the reaction is the result of the structure of the chiral catalyst. The chirality of the catalyst passes on a stereochemical preference to the products, with the result that one of the possible chiral products is formed preferentially in the reaction. Because many chemical processes in living organisms are dependent on asymmetric molecules, the ability to synthesize asymmetric molecules in the laboratory has enormous importance in the production of medicines. L-dopa (where the "L-" stands for levorotatory, i.e., rotating the plane of polarized light to the left) is one such medicine. Jeremy Knowles used a solution of a rhodium complex as a catalyst to achieve chiral hydrogenation, the asymmetric addition of a hydrogen molecule across a double bond. This led to the synthesis of L-dopa, which is sometimes used in trying to moderate the effect of diseases of the nervous system, including Parkinson's disease. Such diseases involve a deficiency of a particular neurotransmitter, dopamine, and L-dopa can be converted to dopamine in the brain.

Chemists have tackled many pharmaceutical organic syntheses, some particularly challenging. One was the synthesis in 1994 of *taxol,*† an anticancer drug that occurs naturally in the Pacific yew tree. The worldwide supply of Pacific yews is limited, and chemists were eager to synthesize an antitumor substance that was one of the most promising tools in treating breast, ovarian, and lung cancers. Two research teams succeeded in the synthesis. The one led by Robert Holton in Florida built the molecule step by step, while the other team, led by K. C. Nicolaou and his colleagues in California, made two large parts of taxol separately and then combined them. It is an unusual molecule, including four rings, and it was very hard to build.

---

*A mixture of equal amounts of left- and right-handed forms of an optically active compound is called a *racemic* mixture.

†Taxol is now a registered trademark of the Bristol-Myers Squibb Company.

The Buckyball and Other Fullerenes

The Royal Swedish Academy of Science's press release for the Nobel Prize in Chemistry in 1996 announced: "The discovery of carbon atoms bound in the form of a ball is rewarded." Buckminsterfullerene (C_{60}), the "buckyball," has the same structure as the geodesic dome designed by Buckminster Fuller for the 1996 Montreal World Exhibition. It also has exactly the same structure as a European soccer ball.

Until 1985, when the fullerenes, new forms of the element carbon, were discovered, chemists knew two kinds of graphite, two kinds of diamond, and two other forms (chaoit and carbon [VI]). Fullerenes were produced by condensing vaporized carbon in an atmosphere of inert gas. C_{60} turns out to be very stable, but it is only one of many new clusters of carbon atoms.

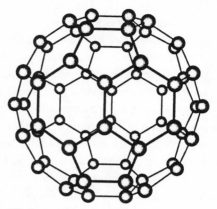

Work on carbon cluster formation may help to show how long carbon-chain molecules may be formed in stellar atmospheres.

■ Quote from *http://www.nobel.se/announcement-96/chemistry96.html*. Drawing in Philip Ball, *Designing the Molecular World* (copyright © 1994 by Princeton University Press).

One of the most elegant classes of molecules is that of the *fullerenes,* which are carbon compounds in the form of hollow spheres, constructed of twelve five-sided faces and different numbers of six-sided faces. The smallest fullerene has thirty-two carbon atoms; the larger ones have several hundred carbon atoms. The first fullerene was discovered in 1985, by two Americans, Richard Smalley (1943–) and Robert Curl (1933–), and an English chemist, Harry Kroto (1939–). The fullerene with sixty carbon atoms, C_{60}, has a structure similar to the geodesic dome invented by the architect Buckminster Fuller. In a whimsical tribute, the whole class of substances was named after the American architect, and his whole name was used for C_{60}, buckminsterfullerene, or, as it is cheerfully known, the buckyball. Fullerenes are stable and can trap other atoms or small molecules inside their spheres. We have scarcely begun to discover their potential uses.

If the fullerenes are among the most elegant and amusing of molecules, there is for most people no question about what is the most important molecular structure ever discovered.* It is the *double helix,* discovered in 1953 by

*One other contender is the benzene molecule.

James Watson (1928–) and Francis Crick (1916–). Watson, then a young American student of biochemistry, came to Cambridge University, where Crick was investigating protein structures, using principally the techniques of X-ray diffraction analysis, that is, the analysis of the patterns produced when X-rays are passed through and scattered by crystallized proteins. Crick wanted to understand the biochemistry of life. Watson urged that it was crucial to discover the structure of the hereditary material, DNA, before they could understand how it worked. The ambitious pair brilliantly and eagerly used the previous work of other scientists, including Linus Pauling, and the crystallographers Rosalind Franklin (1920–58) and Maurice Wilkins (1916–).

In just two years from their first meeting, Crick and Watson came up with their double helix, a kind of double corkscrew. DNA has two strands or chains, twisting and spiraling together in a right-handed helix. Each strand is built up of molecules called nucleotides, of which some of the key components are chemical bases containing hydrogen and nitrogen. There are just four of these bases in DNA, adenine, thymine, guanine, and cytosine, usually referred to by their first letters, A, T, G, and C. The two strands are held together by the relatively weak bonds between a hydrogen atom (positive) of one base with a nitrogen atom of its opposite number in the other strand. A single hydrogen atom thus appears to be bonded to two separate atoms. Such bonds are called *hydrogen bonds*. Geometry and hydrogen bonding cooperate to make the spirals fit and hold together. When the strands become separated, new complementary strands form along each of the original two; here is the basis for replication.

The Chemistry of Life, and the Chemistry of the Environment

Genes and chromosomes, the biological units of heredity, are also chemical units. The strands of DNA carry the genetic code, the information needed for the transmission of heritable features, in their sequence or arrangement of nucleotides. This is the basis for the transmission of chemical information and biological characteristics from one generation to the next.

Almost all organisms use the same genetic code, with the same bases A, T, G, and C. But when it comes to the higher animals, no two organisms, with the exception of identical twins, have the same coded information, the same DNA. That is why DNA analysis has become such a powerful forensic tool. DNA is also a marker of evolutionary descent. The analysis of DNA from Neanderthal tissue in 1987 was conclusive proof that Neanderthals were not ancestors of today's humans, *Homo sapiens*. DNA and related molecules have be-

The Discoverers of the Double Helix

The discovery of the double helix is always associated with two names, Francis Crick and James Watson, shown here (Crick at right) with their model in May 1953, at Cavendish Laboratory, in Cambridge, England. The Nobel Prize in Physiology or Medicine in 1962 was to Crick, Watson, and a third researcher, Maurice Wilkins. Wilkins investigated samples of DNA by X-ray crystallography, and his results indicated that very long chain molecules of DNA were arranged in the form of a double helix. Watson and Crick showed how the organic bases were paired in the intertwined spirals and why this was important as a key to replication.

In the presentation speech for their joint prize, Professor A. Engström stated: "Today no one can really ascertain the consequences of this new exact knowledge of the mechanisms of heredity. We can foresee new possibilities to conquer disease and to gain better knowledge of the interaction of heredity and environment and a greater un-

derstanding for the mechanisms of the origins of life. In whatever directions we look we see new vistas." He was quite right, and the medical, social, and economic implications of that discovery are still being explored and applied at an accelerating rate.

■ Quote from *http://mirror.nobel.ki.se/laureates/medicine-1962-press.html*. Photo used with the permission of James Watson and of Antony Barrington Brown, the photographer.

come so splendidly successful in providing a chemical route into understanding how organisms function, and so promising to many in tracing the history of life on earth, that extravagant claims have become the norm. Thus, for example, John Maddox, for many years editor of the internationally influential journal *Nature,* has made large claims for biochemistry: "The program to understand the origin of life on Earth is thus neither more nor less than an effort to write the natural history of the biochemistry that sustains all life even now." He likewise claimed "the structure of DNA provides not just an understanding of the mechanism of inheritance and of the origins of Darwinian varia-

The DNA Molecule

The top part of this diagram shows a very small section of the interlinked spirals of the double helix, and the horizontal lines indicate pairs of bases linked by hydrogen bonds. This base pairing is shown more clearly in the lower part of the diagram, where the smaller base molecules are cytosine and thymine, and the larger ones are adenine and guanine. Besides these bases, each spiral or helix contains a sugar and a phosphate, repeated all the way along the chain, while the sequence of bases varies such that for species that reproduce sexually, it is unique to every individual (with the exception of identical twins, who share the same DNA).

DNA serves as a blueprint for the individual, controlling, among other aspects, the production of highly specific proteins. In sexual reproduction, the DNA of the offspring is derived from the DNA of both parents, and so serves as the material basis of inheritance and of evolution. Cloning experiments, from Dolly the sheep onward; genetically modified crops; the possible treatment of inherited diseases—in these and other ways, an understanding of the chemical basis of reproduction and in-

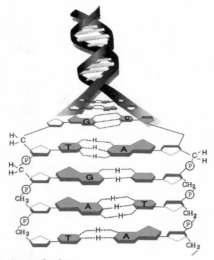

A = adenine
G = guanine
C = cytosine
T = thymine
P = phosphate group

heritance opens "the most spectacular possibilities for the unravelling of the details of the control and transfer of genetic information."

■ Diagram: Tania Hill. Quotation from Professor A. Engström's 1962 presentation speech for the Nobel Prize in Chemistry, *http://mirror.nobel.ki.se/laureates/medicine -1962-press.html.*

tion, but seems to have made it possible to answer any question about the mechanism of life."* DNA may not take us quite as far as Maddox believes. But it assuredly helps us understand how organisms survive, and how they produce the next generation.

One of the most labor intensive, costly, and ambitious projects of modern science is the *human genome* project. The genome of a species is its characteristic set of genes, which makes it what it is. It is a control center, where in-

*John Maddox, *What Remains to Be Discovered: Mapping the Secrets of the Universe, the Origins of Life, and the Future of the Human Race* (New York: Free Press, 1998), 146, 195.

structions for development are stored and from whence they are sent out. It specifies all the chemical components of cells; and when cells divide, the genome itself is replicated. Genes are units of heredity, sequences of nucleotides in the DNA of a species. There are around 100,000 genes in human DNA, and around 3 billion base pairs, so identifying them and determining their sequence is a gigantic project, feasible only with vast funds and huge numbers of research scientists. Because drug companies as well as governments have decided that information about the human genome is worth having, both the funds and the researchers have been made available. But it is still a huge task. Other simpler species have simpler genomes. In 1996 chemists worked out the complete genome of one species of bacteria, which are very much simpler organisms than we are. The human genome project, begun in 1990, was originally planned to last fifteen years, but rapid improvements in technology have accelerated developments, and we now (2001) have a working draft.

Some human diseases can be traced to a single gene, and in those cases it should be possible to diagnose inherited diseases in advance. It may one day become possible to replace "defective" genes with genes carrying healthy or desirable traits. But not all inherited diseases will be easily identified and avoided. In many instances, there are several genes that work together to produce or invite disease, and discovering these combinations will be at best difficult and may be elusive for a very long time; the role of environmental factors also complicates matters. Where human health will benefit, and has indeed already begun to benefit, is in using our knowledge of the human genome to discover the way that diseases are caused.

For example, some individuals have a gene that causes their bodies to produce excessive amounts of cholesterol, and this greatly increases their risk of heart attacks in middle age. In 1985 Joseph Goldstein and Michael Brown received the Nobel Prize in medicine for their work at the University of Texas on regulating cholesterol metabolism. Once the production of cholesterol was understood, it became desirable to produce a drug that would inhibit the action of the gene responsible for excessive cholesterol production. Such drugs, "statins," have now been produced, and they save lives. Atorvastatin (Lipitol), introduced in 1997, is an effective member of this class. Looking ahead, we know that inherited muscular dystrophy is genetically caused; its victims die from one kind or another of muscle failure. But there is also a gene responsible for regenerating atrophied muscle, and current research in Oxford University and elsewhere is directed toward finding a chemical trigger for this gene. If such a trigger is found, and is determined to be otherwise benign, then we shall have a cure for some forms of muscular dystrophy.

The treatment of disease has much to gain from research stimulated by the human genome project. Of course there were many ways in which chemistry contributed to medicine before anyone thought of solving the genome. Chemical medicine goes back centuries, to Paracelsus and even much earlier. It really took off in the twentieth century. The discovery of the sulfonamide drugs was a thoroughly international one, starting from work in 1932 by German biochemist Gerhard Domagk (1895–1964), followed by French work, and culminating in 1936 with the research of Leonard Colebrook (1883–1967). Sulfonamides as a group are powerful antibacterial agents and greatly reduced deaths from infected wounds and injuries. Penicillin, the first of the modern range of antibiotics, was synthesized not by chemists but by nature. Alexander Fleming first realized its significance as an antibiotic in 1928, but it took another ten years for Howard Florey and colleagues at Oxford to complete Fleming's work. Thus synthetic antibacterial agents, the sulfonamides, and natural organic antibiotics were applied in medicine at around the same time, and both became important during World War II. The range of synthetic drugs now available in medicine is startling, and their synthesis and sale is very big business.

The line between conventional chemical synthesis and the production of chemicals by living agents is hard to draw. The vitamins that we buy at the pharmacy or health food store may be made by chemists or, as they increasingly are, by carefully designed bacteria. The products in both cases are the same: vitamin C produced by bacteria is the same as vitamin C produced by a chemist using traditional synthetic methods. In a sense, bacteria can be regarded as tools for the chemist to use, although not everyone agrees that using bacteria is part of chemistry. The crucial material for bacteria and other living things is DNA, a chemical embodiment of the genetic code. Genes, in that code, can be thought of as sentences in a very long book. And, just as a written document can be edited by deletions, insertions, and cutting and pasting, so DNA can be edited by deleting, inserting, or rearranging genes. Such activities go under a variety of names, including gene splicing, gene transfer, gene therapy, and genetic engineering. Most chemists regard them as belonging to biology rather than to chemistry; disciplinary boundaries are sometimes blurred, and disciplines on both sides of the boundary contribute to its definition. Both sides of the boundary are important in modern agriculture.

Growing populations need ever more food, produced by ever more efficient agriculture. We demand high crop yields and low prices. Use of fertilizers and pesticides, together with selective breeding of productive and pest-resistant varieties of crops, help to produce increased yields. But some 40 percent of the world's crops are still lost to pests and disease before they come to market. Conventional agriculture relies on large quantities of herbicides and fertilizers, but they have their environmental downside, as Rachel Carson powerfully argued in *Silent Spring* in 1962. Genetically modified foods, the subject of lively debate, may be able to avoid some of these problems. If you can identify the genes that confer pest-resistance in wild plants and then splice them into crops, you get higher yields and may be able to get those yields using less pesticide.

Apart from the mainly organic chemistry involved in crop production, there is the mainly inorganic chemistry used in many industries. And there too, chemistry can help us to understand and resolve problems produced by industry, which may or may not be primarily chemical industry. Acid rain is produced mainly by the combustion of sulfur when cheap coal is burned in homes or power stations, or when heavy oils with high sulfur content are refined. The effects on forests, lakes, and the creatures that live in and around them (including people) have been thoroughly documented. It is possible to install scrubbers to remove the sulfur from emissions at oil refineries, and, as some oil companies have found, this can pay for itself in a few years. Sulfur as a byproduct yields a lot of sulfuric acid, which is still a very important industrial chemical and thus a significant source of revenue. Cleaning up some industries that leak toxic chemicals is less easy, because pollutants such as PCBs linger in the environment. Unfortunately, when much of the pollution was being created by improper disposal, the dangers posed by many pollutants were unknown. Chemistry can often help with the cleanup. Now that we have a fuller understanding of environmental consequences, we need political as well as scientific tools to achieve any improvements.

Chain Reactions in the Ozone Layer, and Smog in the City

From the 1950s into the 1980s, the refrigerant most widely used in freezers, air conditioners, and refrigerators was ammonia. If leaks developed in the cooling system, ammonia was released. It is a choking, highly reactive, and potentially harmful gas. Chemists looked for a replacement that would be safer to use. They hit on chlorofluorocarbons (also called CFCs, or Freons), stable compounds of carbon, fluorine, and chlorine which were not toxic, not inflammable, and had the right physical properties to act as refrigerants. They were also soon used as the propellant in spray cans. Unfortunately, as chemists came to understand in the 1970s, their use had an unintended and unexpected consequence. Freons accumulate in the stratosphere, one of the upper reaches of the earth's atmosphere, and there are decomposed by ultraviolet radiation from the sun. In decomposing, Freon molecules release individual chlorine atoms, which then combine with ozone, a triatomic molecular form of oxygen, O_3, which is present in the stratosphere. Single chlorine atoms have no net electrical charge (unlike ions) and are highly reactive, and in that highly reactive state are known as *free radicals*. They break down the molecules of ozone and are then given up by the oxygen atoms to recombine with another ozone molecule, and the cycle of ozone destruction begins again. This cycle can be renewed very many times, in what is known as a *chain reaction*. Such reactions were first explained (although not under this name) by Max Bodenstein (1871–1942), a German photochemist and chemical kineticist, who was Nernst's successor at the University of Berlin.

In the course of a chain reaction, a single Freon molecule can cause the decomposition of millions of ozone molecules, and a small amount of Freon can have a long-lasting effect. Stratospheric ozone is a major protection against ultraviolet radiation from the sun, and we have learned that a small decrease in the concentration of stratospheric ozone leads to harmful environmental consequences. By the 1990s, the public was thoroughly informed about the relation between excess exposure to ultraviolet radiation and increased dangers of skin cancer in humans and genetic mutations in other organisms. The connection between the decrease in ozone and the increase in cancer is clear and undisputed. In the late 1970s, Canada, the United States, and the Scandinavian countries banned the use of CFCs in spray cans, and in the early 1990s around a hundred countries agreed to phase out the production of CFCs altogether before the decade was out. That does not take care of the CFCs already in the stratosphere, and their effect will last through much of the twenty-first century.

Lots of ozone in the stratosphere is good; loss of ozone is bad. Lower down, where we live, ozone is simply bad. It ranks as highly toxic. The main source of it in cities is from automobile exhaust, which contains oxides of nitrogen and hydrocarbons. Sunlight acts on this mixture, and ozone is one product; nitrogen dioxide, a toxic and corrosive gas, is also produced by the action of sunlight on nitrogen oxide. Sometimes, especially in summer, there are *atmospheric inversions,* when a warm layer of air sits over a cooler mass below it and acts like a lid on a pot, keeping everything inside the pot. When there is an inversion in cities with a lot of automobile emissions, for example in Los Angeles or in Mexico City, smog is the result. The nitrogen dioxide gives a brownish color to the air, and the mix of nitrogen dioxide and ozone causes respiratory problems, itching eyes and throat, and damage to plants as well as people. Emission controls, cleaner fuels, more efficient automobile engines, and reduced use of personal automobiles are four of the strategies now in use to reduce smog. Chemists have made significant contributions to the development of cleaner fuels, and catalytic converters in automobiles break down some of the worst emissions. Air quality in Los Angeles in 2000 was considerably better than it was in 1980, although there were more cars, and there is still a way to go. Meanwhile, until there are significant changes in technology and lifestyle, there are many sunny days in cities when there is an air advisory warning and it is best to stay indoors.

Chemistry is everywhere, even in industries that do not seem chemical. Making a single electronic microcircuit, for example, involves nearly one hundred different chemical processes. The air we breathe, the water we drink, the food we eat, and even the sunlight that we used to bask in, all are affected by chemical processes that we understand better thanks to new research. We depend on synthetic materials in our homes and workplaces, and even in our clothing. Our society is built in large part on plastics and semiconductors. Many of the problems that we confront have causes that chemistry can elucidate. Many of the remedies that we apply to these problems are chemical remedies, which often, as in the agreement to phase out CFCs, involve political solutions too. Whether we like it or not, the present scale of food production around the globe would be impossible without extensive use of chemical aids, in the form of pesticides and fertilizers.

As we have seen, the boundaries of chemical science are fluid and changing, more rapidly and on more fronts now than ever before. There is no reason to assume that this fluidity will come to a sudden stop. The history of chemistry is the story of a science in perpetual flux, and we do not know where it will take us next. Even when we feel confident that we know what lies ahead, we cannot be sure what shape the science will take even a little way down the road. We can have reasonable confidence that some of the questions being asked today will be answered, but we should be even more confident that the twenty-first century will see chemists posing questions and developing fields that we have not yet imagined. Today's chemistry, "modern" chemistry, will by then be part of tomorrow's history.

Suggested Further Reading

There are several more detailed and longer histories of chemistry, including W. H. Brock, *The Norton History of Chemistry* (New York: Norton, 1993), Bernadette Bensaude-Vincent and Isabelle Stengers, *A History of Chemistry* (Cambridge, Mass.: Harvard University Press, 1996), and, for reference rather than for reading, J. R. Partington, *A History of Chemistry,* 4 vols. (London: Macmillan, 1962–70).

1 First Steps: From Alchemy to Chemistry?

Three books offer both a general discussion of the history of alchemy and a detailed study of the alchemy of Robert Boyle, George Starkey, and Isaac Newton: Lawrence M. Principe, *The Aspiring Adept: Robert Boyle and His Alchemical Quest* (Princeton: Princeton University Press, 1998); William R. Newman, *Gehennical Fire: The Lives of George Starkey, an American Alchemist in the Scientific Revolution* (Cambridge, Mass.: Harvard University Press, 1994); and B. J. T. Dobbs, *The Janus Faces of Genius: The Role of Alchemy in Newton's Thought* (Cambridge: Cambridge University Press, 1991). There is useful discussion of Paracelsus's alchemy in Piyo Rattansi and Antonio Clericuzio, eds., *Alchemy and Chemistry in the Sixteenth and Seventeenth Centuries* (Dordrecht: Kluwer, 1994). Chaucer's "Canon's Yeoman's Tale" may be read in any good edition of his *Canterbury Tales,* or in modern translation, Geoffrey Chaucer, *The Canterbury Tales,* trans. David Wright (Oxford: Oxford University Press, 1985). Brock, *The Norton History of Chemistry,* discusses the modern achievement of transmutation.

2 Robert Boyle: Chemistry and Experiment

Marie Boas Hall, *Robert Boyle and Seventeenth-Century Chemistry* (Cambridge: Cambridge University Press, 1958), presents Boyle as a corpuscular philosopher. Steven Shapin and Simon Shaffer, *Leviathan and the Air-Pump* (Princeton: Princeton University Press, 1985), discuss Boyle's work with the air pump in its social and rhetorical contexts. More recently, Rose-Mary Sargent, *The Diffi-*

dent Naturalist: Robert Boyle and the Philosophy of Experiment (Chicago: University of Chicago Press, 1995), has given a more careful examination of Boyle's methods; the account of Boyle's philosophy of experiment in this chapter is directly based on Sargent's work. Lawrence M. Principe, *The Aspiring Adept: Robert Boyle and His Alchemical Quest* (Princeton: Princeton University Press, 1998), gives the fullest and best account of Boyle's alchemy and uses the term *chymistry.* W. R. Newman and L. M. Principe, "Alchemy *vs.* Chemistry: The Etymological Origins of a Historiographic Mistake," *Early Science and Medicine* 3 (1998): 32–65, shows how mistakes in interpreting labels can distort the history of chemistry. William R. Newman, *Gehennical Fire: The Lives of George Starkey, an American Alchemist in the Scientific Revolution* (Cambridge, Mass.: Harvard University Press, 1994), discusses the work of one of the main contemporary influences on Boyle.

3 A German Story: What Burns, and How

For Paracelsus, see Walter Pagel, *Paracelsus: An Introduction to Philosophical Medicine in the Era of the Renaissance* (Basel: Karger, 1958). Allen Debus, *The English Paracelsians* (New York: Franklin Watts, 1966), gives a good account of Paracelsian medicine, and the same author, in *The Chemical Philosophy: Paracelsian Science and Medicine in the Sixteenth and Seventeenth Centuries,* 2 vols. (New York: Science History Publications, 1977), ties iatrochemistry to the chemistry of Becher and Stahl. A fine narrative that bridges Newton to Stahl and Boerhaave, unfortunately still not translated into English, is Hélène Metzger, *Newton, Stahl, Boerhaave et la doctrine chimique* (Paris: Librairie Félix Alcan, 1930). The comparison of Stahl and Rouelle is based on Bernadette Bensaude-Vincent and Isabelle Stengers, *A History of Chemistry* (Cambridge, Mass.: Harvard University Press, 1996). Becher's career is well described in Pamela H. Smith, *The Business of Alchemy: Science and Culture in the Holy Roman Empire* (Princeton: Princeton University Press, 1994). The various strands of the phlogiston theory are discussed in J. R. Partington and D. McKie, "Historical Studies on the Phlogiston Theory," pts. I–IV, *Annals of Science* 2 (1937): 361–404; 3 (1938): 1–58; 3 (1938): 337–71; 5 (1939): 113–49.

4 An Enlightened Discipline: Chemistry as Science and Craft

The most insightful account of eighteenth-century chemistry before Lavoisier is F. L. Holmes, *Eighteenth-Century Chemistry as an Investigative Enterprise* (Berkeley: Office for the History of Science and Technology, University of California, 1989); I have based my account of the chemistry of salts on Holmes's account in chap. 2. For an account of the history of the language of chemistry, see Maurice P. Crosland, *Historical Studies in the Language of Chemistry* (London: Heinemann, 1962). The most complete history of chemical affinity (not yet translated into English) is Michelle Goupil, *Du flou au clair? Histoire de l'affinité chimique de Cardan à Prigogine* (Paris: Comité des Travaux Historiques et Scientifiques, 1991). For Scottish chemistry and its context, see A. L. Donovan, *Philosophical Chemistry in the Scottish Enlightenment: The Doctrines and Discoveries of William Cullen and Joseph Black* (Edinburgh: Edinburgh University Press, 1975), and Archibald and Nan Clow, *The Chemical Revolution* (London: Batchworth Press, 1952).

5 Different Kinds of Air

An excellent introductory essay to eighteenth-century chemistry and the chemical revolution is Maurice P. Crosland, "Chemistry and the Chemical Revolution," in G. S. Rousseau and R. Porter, eds., *The Ferment of Knowledge: Studies in the Historiography of Eighteenth-Century Science* (Cambridge: Cambridge University Press, 1980). Crosland discusses early instruments for the handling of gases in chap. 4 of F. L. Holmes and T. H. Levere, eds., *Instruments and Experimentation in the History of Chemistry* (Cambridge, Mass.: MIT Press, 2000), which I have drawn on freely. Levere discusses eudiometry, the measurement of the goodness of gases, in chap. 5 of the same book, and Holmes discusses Lavoisier's research apparatus in chap. 6. Joseph Black, *Experiments upon Magnesia Alba* (1755), has been reprinted as *Alembic Club Reprint, No. 1* (Edinburgh: Livingstone, 1963). Joseph Priestley appears engagingly in Robert E. Schofield, ed., *A Scientific Autobiography of Joseph Priestley (1783–1804)* (Cambridge, Mass.: MIT Press, 1966). Schofield has also written a fine account of Priestley's first forty years, *The Enlightenment of Joseph Priestley: A Study of His Life and Work from 1733–1773* (Philadelphia: Pennsylvania State University Press, 1997). The literature on Lavoisier and the chemical revolution is enormous. The best introduction remains Henry Guerlac, *Antoine-Laurent Lavoisier: Chemist and Revolutionary* (Scribner's: New York, 1975); a good recent biography is Jean-Pierre Poirier, *Lavoisier: Chemist, Biologist, Economist* (Philadelphia: University of Pennsylvania Press, 1996). Two excellent detailed studies of Lavoisier's laboratory practice are F. L. Holmes, *Lavoisier and the Chemistry of Life: An Exploration of Scientific Creativity* (Madison: University of Wisconsin Press, 1985), and, by the same author, *Antoine Lavoisier—The Next Crucial Year: Or, The Sources of His Quantitative Method in Chemistry* (Princeton: Princeton University Press, 1998). Lavoisier's book, *Elements of Chemistry*, trans. R. Kerr (Edinburgh, 1790; actually published in 1789), has been reprinted (New York: Dover, 1965).

6 Theory and Practice: The Tools of Revolution

Several of the key sources are given at the end of the listing for chapter 5, including Guerlac's and Poirier's biographies of Lavoisier. F. L. Holmes, *Antoine Lavoisier—The Next Crucial Year: Or, The Sources of His Quantitative Method in Chemistry* (Princeton: Princeton University Press, 1998), as the subtitle promises, gives an excellent account of Lavoisier's quantitative approach. Two of Lavoisier's key instruments are described in paper V, "Balance and Gasometer in Lavoisier's Chemical Revolution," in T. H. Levere, *Chemists and Chemistry in Nature and Society 1770–1878* (Aldershot, Hampshire, and Brookfield, Vt.: Variorum, 1994). There are fine essays on the tools and methods of eighteenth-century analysis, including the blowpipe, in David Oldroyd, *Sciences of the Earth: Studies in the History of Mineralogy and Geology* (Aldershot, Hampshire, and Brookfield, Vt.: Variorum, 1998). See also Brian P. Dolan, "Blowpipes and Batteries: Humphry Davy, Edward Daniel Clark, and Experimental Chemistry in Early Nineteenth-Century Britain," *Ambix* 15 (1998): 137–62, and "Transferring Skill: Blowpipe Analysis in Sweden and England, 1750–1850," in Brian P. Dolan, ed., *Science Unbound: Geography, Space and Discipline* (Umeå: Umeå University Press, 1998), 91–125. Lavoisier's collaboration with Laplace is presented in Lavoisier and Laplace, *Memoir on Heat Read to the Royal Academy of Sciences 28 June 1783,* trans. and ed. Henry Guerlac (New York: Neal Watson Academic Publications, 1982).

7 Atoms and Elements

For Lavoisier, see sources given at the end of the listing for chapter 6. For Dalton, the most accessible account is Elizabeth C. Patterson, *John Dalton and the Atomic Theory: The Biography of a Natural Philosopher* (New York: Doubleday, 1970). David M. Knight, *Atoms and Elements: A Study of Theories of Matter in England in the Nineteenth Century* (London: Hutchinson, 1967), has a good discussion about the British response to Dalton. A more international approach is Alan J. Rocke, *Chemical Atomism in the Nineteenth Century from Dalton to Cannizzaro* (Columbus: Ohio State University Press, 1984). For Davy, see Harold Hartley, *Humphry Davy* (London: Nelson, 1966), and David M. Knight, *Humphry Davy: Science and Power* (Oxford: Blackwell, 1992). Berzelius is discussed in Rocke, *Chemical Atomism,* and in Evan M. Melhado, *Jacob Berzelius: The Emergence of His Chemical System* (Madison: University of Wisconsin Press, 1981). Trevor H. Levere, *Affinity and Matter: Elements of Chemical Philosophy 1800–1865* (Oxford: Clarendon Press, 1971; reprint, Yverdon and Langhorne: Gordon and Breach, 1993), discusses both Davy and Berzelius. There is a fine account of Berzelius's methodology in John Hedley Brooke,

Thinking About Matter: Studies in the History of Chemical Philosophy (Aldershot, Hampshire, and Brookfield, Vt.: Variorum, 1995).

8 The Rise of Organic Chemistry

F. L. Holmes, *Eighteenth-Century Chemistry as an Investigative Enterprise* (Berkeley: Office for the History of Science and Technology, University of California, 1989), has an excellent chapter on eighteenth-century vegetable chemistry. For Lavoisier's research in organic chemistry, see the same author's *Lavoisier and the Chemistry of Life: An Exploration of Scientific Creativity* (Madison: University of Wisconsin Press, 1985). The background to the arguments in this chapter is explored in T. H. Levere, *Affinity and Matter: Elements of Chemical Philosophy 1800–1865* (Oxford: Clarendon Press, 1971; reprint, Yverdon and Langhorne: Gordon and Breach, 1993). The methodological debates are explored in John Hedley Brooke, *Thinking About Matter: Studies in the History of Chemical Philosophy* (Aldershot, Hampshire, and Brookfield, Vt.: Variorum, 1995). Liebig and Wöhler's paper on the benzoyl radical can be read in translation in O. T. Benfey, *Classics in the Theory of Chemical Combination* (New York: Dover, 1963), along with key writings by Laurent and Gerhardt.

9 Atomic Weights Revisited

Papers by Prout, Stas, and Marignac are available in English translation in *Prout's Hypothesis, Alembic Club Reprints, No. 20* (Edinburgh: The Alembic Club, 1932). There is a fine account of the history of Prout's hypothesis and its modern successors in W. H. Brock, *From Protyle to Proton: William Prout and the Nature of Matter 1785–1985* (Bristol and Boston: Adam Hilger, 1985). There is a useful essay on the fate of Avogadro's hypothesis in John Hedley Brooke, *Thinking About Matter: Studies in the History of Chemical Philosophy* (Aldershot, Hampshire, and Brookfield, Vt.: Variorum, 1995). Mendeléev's key papers, as well as those of others who explored the relationship between atomic weights and chemical properties, are reproduced in English in David M. Knight, ed., *Classical Scientific Papers: Chemistry, Second Series: Papers on the Nature and Arrangement of the Chemical Elements* (London: Mills and Boon, 1970).

10 The Birth of the Teaching-Research Laboratory

Maurice Crosland, *The Society of Arcueil: A View of French Science at the Time of Napoleon* (London: Heinemann, 1971), gives an excellent account of Gay-Lussac and Berthollet's institution. Besides W. H. Brock's biography, *Justus von Liebig: The Chemical Gatekeeper* (Cambridge: Cambridge University Press, 1997), I have used three important articles on Liebig and his research school and teaching laboratory: J. B. Morrell, "The Chemist Breeders: The Research Schools of Liebig and Thomas Thomson," *Ambix* 19 (1972): 1–46; Joseph S. Fruton, "The Liebig Research Group—A Reappraisal," *Proceedings of the American Philosophical Society* 132 (1988): 1–49; and F. L. Holmes, "The Complementarity of Teaching and Research in Liebig's Laboratory," *Osiris* 5 (1989): 121–64. Alan J. Rocke has written an admirable account of the use of Liebig's potash apparatus, "Organic Analysis in Comparative Perspective: Liebig, Dumas, and Berzelius, 1811–1837," in F. L. Holmes and T. H. Levere, eds., *Apparatus and Experimentation in the History of Chemistry* (Cambridge, Mass.: MIT Press, 2000), 273–310. See also F. L. Holmes, "Justus Liebig and the Construction of Organic Chemistry," in Seymour Mauskopf, ed., *Chemical Sciences in the Modern World* (Philadelphia: University of Pennsylvania Press, 1993). For chemical industry, see L. F. Haber, *The Chemical Industry during the Nineteenth Century* (Oxford: Clarendon Press, 1958). For an account of the place of organic chemistry in German university laboratories, see Alan J. Rocke, *The Quiet Revolution: Hermann Kolbe and the Science of Organic Chemistry* (Berkeley and Los Angeles: University of California Press, 1993). The situation in France is well described in A. J. Rocke, *Nationalizing Science: Adolphe Wurtz and the Battle for French Chemistry* (Cambridge, Mass.: MIT Press, 2001).

11 Atoms in Space

For Frankland, see C. A. Russell, *Edward Frankland: Chemistry, Controversy and Conspiracy in Victorian England* (Cambridge: Cambridge University Press, 1996). Several of the key papers discussed in this chapter are reprinted in translation in O. T. Benfey, *Classics in the Theory of Chemical Combination* (New York: Dover, 1963). The best translation of Van't Hoff's paper is in Peter J. Ramberg and Geert J. Somsen, "The Young J. H. van't Hoff: The Background to the Publication of His 1874 Pamphlet on the Tetrahedral Carbon Atom, Together with a New English Translation," *Annals of Science* 58 (2001): 51–74. C. A. Russell, *The History of Valency* (Leicester: Leicester University Press, 1971), pts. 1–3, discusses bonds, formulas, the theory of structure, and, of course, valence. There are useful essays in two centenary volumes: O. T. Benfey, ed., *Kekulé Centennial* (Washington, D.C.: American Chemical Society, 1966), and O. Bertrand Ramsay, ed., *van't Hoff–Le Bel Centennial* (Washington, D.C.: American Chemical Society, 1975). For coordination chemistry, see George B. Kauffman, *Alfred Werner—Founder of Coordination Chemistry* (New York: Springer, 1966), and *Classics in Coordination Chemistry*, 3 vols. (New York: Dover, 1968 and 1976). Russell, *The History of Valency*, 213–23, has an elegant analysis of the thinking in Werner's key papers. For the Australian contribution to the twentieth-century revival of inorganic chemistry, see Ronald Nyholm, *Journal of Chemical Education* 34 (1957): 166–69, and W. H. Brock, *The Norton History of Chemistry* (New York: Norton, 1993), chap. 15; in chap. 16, Brock discusses organic synthesis and organic chemical industry.

12 Physical Chemistry: A Discipline Comes of Age

The pioneering work on the theory of heat engines was by Sadi Carnot in 1824, translated into English as *Reflections on the Motive Power of Heat* (1897), and now available in paperback (New York: Dover, 1960). A fine introduction to the range of physical chemistry and its history is Keith J. Laidler, *The World of Physical Chemistry* (Oxford: Oxford University Press, 1993). I have used this book extensively in this chapter. The mathematical parts of the book will be hard for some readers, but even if they skip the mathematics, they will still learn a lot. A useful study of the early years of physical chemistry is John W. Servos, *Physical Chemistry from Ostwald to Pauling: The Making of a Science in America* (Princeton: Princeton University Press, 1990); chap. 1 is a good introduction to the beginnings of the discipline. Physical chemistry as a discipline is also discussed in Mary Jo Nye, *From Chemical Philosophy to Theoretical Chemistry: Dynamics of Matter and Dynamics of Disciplines 1800–1950* (Berkeley and Los Angeles: University of California Press, 1993). The formulation of the three laws of thermodynamics in terms of a game of chance comes from Steven J. Groák, *The Idea of Building: Thought and Action in the Design and Production of Buildings* (London: E & FN Spon, 1992), 45. There is an admirable study of Nernst and physical chemistry in Diana Kormos Barkan, *Walther Nernst and the Transition to Modern Physical Science* (Cambridge: Cambridge University Press, 1999). Barkan also discusses the origins of the discipline of physical chemistry in "A Usable Past: Creating Disciplinary Space for Physical Chemistry," in Mary Jo Nye, Joan Richards, and Roger Stuewer, eds., *The Invention of Physical Science: Intersections of Mathematics, Theology and Natural Philosophy since the Seventeenth Century: Essays in Honor of Erwin N. Hiebert*, Boston Studies in the Philosophy of Science, vol. 139 (Dordrecht: Kluwer, 1992), 175–202.

13 The Nature of the Chemical Bond

The title of this chapter is from Linus Pauling, *The Nature of the Chemical Bond and the Structure of Molecules and Crystals* (Ithaca, N.Y.: Cornell University Press, 1939); there have been several subsequent editions, any of which is well worth consulting. Keith J. Laidler, *The World of Physical Chemistry* (Oxford: Oxford University Press, 1993), has valuable accounts of several of the developments discussed in this chapter. So do the later chapters of C. A. Russell, *The History of Valency* (Leicester: Leicester University Press, 1971). There are some excellent papers on the history of the chemical bond by Robert E. Kohler, Jr., of which the most relevant here is "The Origins of G. N. Lewis's Theory of the Shared Pair Bond," *Historical Studies in the Physical Sciences* 3 (1971): 342–76. For an insider's account of molecular spectroscopy, see G. Herzberg, "Molecular Spectroscopy: A Personal History," *Annual Review of Physical Chemistry* 36 (1985): 1–30. An introductory account of the history of quantum theory may be found in A. Einstein and L. Infeld, *The Evolution of Physics* (New York: Simon and Schuster, 1938; several reprints).

14 Where Now, and Where Next? New Frontiers

For a guide to the new artificial elements, consult Albert Stwertka, *A Guide to the Elements,* rev. ed. (New York: Oxford University Press, 1998). For the new physical chemistry, see Keith J. Laidler, *The World of Physical Chemistry* (Oxford: Oxford University Press, 1993). On buckminsterfullerene, see Hugh Aldersey-Williams, *The Most Beautiful Molecule: The Discovery of the Buckyball* (New York: John Wiley, 1995). For an account of a variety of fascinating molecules, see Philip Ball, *Designing the Molecular World: Chemistry at the Frontier* (Princeton: Princeton University Press, 1994). For the discovery of DNA, there are accounts in two very different styles: James D. Watson, *The Double Helix: A Personal Account of the Discovery of the Structure of DNA* (New York: Athenaeum, 1968), and R. C. Olby, *The Path to the Double Helix* (London: Macmillan: 1974; reprint, New York: Dover, 1994). For relatively modern biochemistry, see Steven Rose with Sarah Bullock, *The Chemistry of Life,* 3rd ed. (London: Penguin Books, 1991). There is much about the role of chemistry in relation to biology in John Maddox, *What Remains to Be Discovered: Mapping the Secrets of the Universe, the Origins of Life, and the Future of the Human Race* (New York: Free Press, 1998). For the history of genetic engineering, see Susan Aldridge, *The Thread of Life: The Story of Genes and Genetic Engineering* (Cambridge: Cambridge University Press, 1996). For the downside of genetically modified crops, see Jane Rissler, *The Ecological Risks of Engineered Crops* (Cambridge, Mass.: MIT Press, 1996). On the atmosphere, smog, and the ozone layer, see John DeWitt Horel and Jack Geisler, *Global Environmental Change: An Atmospheric Perspective* (New York: John Wiley, 1997); Annika Nilsson, *Ultraviolet Reflections: Life under a Thinning Ozone Layer* (New York: John Wiley, 1996); Derek M. Elsom, *Smog Alert: Managing Urban Air* (London: Earthscan Publications, 1996).

Index